PRACTICAL INSECT PEST MANAGEMENT

PRACTICAL INSECT PEST MANAGEMENT

A Self-Instruction Manual

Theo F. Watson / Leon Moore / George W. Ware

THE UNIVERSITY OF ARIZONA

W. H. FREEMAN AND COMPANY
San Francisco

Printed in the United States of America

International Standard Book Number: 0-7167-0558-3

9 8 7 6 5 4 3 2 1

Dedicated to the Memory of

DWIGHT ISELY

(1887 - 1974)

Late Professor of Entomology, Emeritus,
University of Arkansas, Fayetteville,
who developed many of the strategies
on which insect pest management is built.

PREFACE

Insect pest management combines the best of all useful techniques into custom-made insect control systems, and thus exemplifies the methodology of the future. The consequences of an almost total dependence on chemical pest control for the past 30 years have caught up with us. These consequences are (1) pest resistance to many of the pesticides, (2) biomagnification of pesticidal residues through food chains, (3) adverse effects on non-target organisms, (4) persistence of pesticides in the environment, (5) increased crop-production costs as a result of increased costs of pesticides, and (6) an overreaction by a concerned public to extensive amounts of pesticides being released into the environment.

It becomes essential then that every person selling, recommending, or using insecticides in any aspect of agriculture know something about this seemingly "new" art and science, insect pest management, i.e., the practical manipulation of insect (or mite) pest populations using any or all of a number of management methods and based on sound ecological principles.

Most books on insect pest management are essentially theoretical; this book is not. It contains easy-to-understand outlines of insect pest management that can lead those who study it into a comprehensive appreciation of the "thinking-man's" insect control. This is a "how to" book, a detailed explanation of practical, everyday insect pest management.

Our aim is to explain insect pest management to the layman--to give him an appreciation for the present state of this fine art and science. The reader is carried from the 19th century, when insect pest control began, to the mid 1970's when we are "putting it all together."

This book may seem repetitious at times. It hammers on simple basic procedures, and "Repetition is the master of studies," as the old adage goes. It is a textbook of practical formulae for insect pest management.

Because of the magnitude of the subject matter to be covered, no book of this kind could be complete. Instead, it treats only the insect pests of crop plants, and does not deal with household insects or those directly affecting man and other animals. The book is intended to present a comprehensive picture of insect pest management in its present state; it is not an exhaustive study of each component and its application. It is a self-teaching guide for agricultural fieldmen, pesticide salesmen, county agricultural agents, and high school and college students. We hope that it will also be of interest and useful to those preparing for federal or state certification and licensing in the field of pesticide usage.

Many ideas and some data are presented in various sections of the manual without direct citation of sources. Following each section is a listing which acknowledges the contributors whose work or writings were used.

Acknowledgements. During the writing of this text, we have received valuable advice, criticism, and assistance from many people. In particular, we would like to record our gratitude to:

Dr. Charles G. Lincoln, Mr. W. P. Boyer, Dr. Jake R. Phillips, Dr. Arthur J. Mueller, and Dr. Floyd D. Miner of the University of Arkansas for assessment of the original concept and encouragement to proceed; and Dr. Daniel A. Roberts of the University of Florida and Dr. Larry P. Pedigo of the Iowa State University for reviewing the manuscript. Special recognition is given to Mrs. Hazel Tinsley who typed the camera-ready manuscript as if it were her own.

Finally, to Elizabeth, June and Doris, our wives, who provided us with help and encouragement while sacrificing evenings and weekends, permitting us to complete the manuscript, we owe the most special debt of appreciation.

Tucson, Arizona
October, 1975

Theo F. Watson
Leon Moore
George W. Ware

CONTENTS

INSTRUCTIONS FOR USING THE MANUAL

This is a self-teaching manual. Its somewhat unique format is designed to help you use it as an aid for self-instruction. In Unit 1, where the questions begin, use a shield (paper or an index card) to cover the answers while exposing only the question. Write the answers you think are correct, then slide the shield down just enough to reveal the answer. In some sections several questions are presented on a single page.

You may be tempted to peek at the correct response before recording your answer, but this will stop when you realize that the purpose of the book is to teach rather than test. You are the only person who will be checking for errors, and no one cares how many mistakes you make as long as you are convinced at the end that you know what is correct.

The time required to work through the manual will vary with the reader's background and study habits. The average time should be about 16 hours, varying between 12 and 22 hours.

Tests with self-instruction and programmed texts have shown that students who complete the study by writing the answers score higher on tests than those who do not. So if you are to achieve the objectives which the program is designed to help you attain, you should work through the material in the sequence presented, and write your answers.

Insect pest management is a vital subject in the agricultural technology of the last quarter of the 20th century. We know you will enjoy studying this subject in its challenging format and become much better informed as a result.

Good luck!

BEHAVIORAL OBJECTIVES

A self-instruction manual should focus on a particular audience and present clear-cut goals stated in terms of the degree of competence required of the student upon completion of each unit. These goals are known as "behavioral objectives."

The audience for whom this manual was written can be divided into four main groups. First, there are specialists in insect pest management, agricultural-chemical fieldmen, county agricultural agents, and growers, all of whom are primarily interested in the most satisfactory and environmentally sound way of "managing their agricultural insect pests." Second, there are college and advanced high school students who can use this manual as a supplement to assigned textbooks for courses on pest control, plant protection, chemical insect control, biological control, ecology, and certain environmental studies. Third, there are state and federal officials who are responsible for the preparation of study materials for, and the training of, persons seeking licensing and certification as required by recent state and federal legislation. Finally, concerned citizens who seek a greater understanding of the science and technology that undergird our crop-protection practices may find the manual helpful.

Practical Insect Pest Management: A Self Instruction Manual was written to teach and clarify a selected set of objectives the reader should attain after each unit. They constitute a convenient checklist of terminology, concepts, and pest management tactics to be mastered. Instructors may use these objectives as standards for measurement, both for designing course content and for preparing tests.

After he has progressed through this manual and has successfully answered the questions it contains, the reader should be able to do the assigned tasks enumerated below.

Unit 1 THE NATURE OF INSECT PEST MANAGEMENT

a. Distinguish between pest control, integrated control, and insect pest management.
b. Explain the statement, "Pest management is a dynamic system."
c. List the major reasons for emphasizing development of insect pest management.
d. Describe the ultimate goal of insect pest management.
e. Identify the four basic elements of insect pest management.
f. Explain how single-component controls contribute to insect pest management.

g. Discuss briefly several of the classic single-component insect controls.

Unit 2 BASIC ELEMENTS OF INSECT PEST MANAGEMENT

A-Natural Control

a. Define natural control.
b. List the major environmental components involved in natural control.
c. Distinguish between natural control and naturally occurring biological control.
d. Explain the two principal views of natural control.
e. Differentiate between determination and regulation of populations of insect pests.

B-Sampling

a. Explain the importance of sampling to insect pest management.
b. List four ways of sampling crops for the presence of insect pests.
c. Discuss the use of insect traps in insect-pest-management programs.
d. Differentiate between assessment of insect populations and their damage.
e. Explain the value of counting nondestructive insect stages such as eggs.
f. Outline a sampling pattern for an average field.
g. Design a field record form.
h. Relate time and frequency schedules for field sampling to crop and pest.
i. Summarize the training required by field scouts.
j. Delineate the factors that influence sampling.

C-Economic Levels

a. Define economic levels of insect populations.
b. Explain the significance of economic levels to insect pest management.
c. Distinguish between economic levels in terms of crop damage and economic levels in terms of all economic aspects of insect pest management.
d. Discuss the value of an economic level, even though it is continuously changing.

D-Insect Biology and Ecology

a. Explain the importance of knowing the identity of the insect.
b. Discuss the significance of knowing the biology and ecology of most insects in the agroecosystem.
c. Give examples of how biological or behavioral differences in insect pests affect insect-pest-management strategies.

Unit 3 COMPONENTS OF INSECT PEST MANAGEMENT

A-Cultural Control

a. Define cultural control.
b. Describe the essentials of cultural control.
c. List the advantages and disadvantages of cultural control methods.
d. Name the different cultural practices available for coping with pest problems.
e. Discuss the additive effects of several cultural practices used against a single insect pest.

B-Biological Control

a. Define biological control.
b. List the advantages of biological control over other methods.
c. Distinguish between classical and naturally occurring biological control.
d. Explain how farm management practices are important to beneficial insects.

C-Chemical Control

Introduction, Benefits and Consequences

a. Explain the reasons for not returning to insect pest control practices of a generation ago.
b. Trace the history of our dependence on systematic applications of insecticides for insect control.
c. Identify the two characteristics of a chemical necessary for its biomagnification.
d. Explain how insects become resistant to insecticides.
e. Describe the effects of insecticides on nontarget organisms.
f. Explain how treated pest species resurge.

Formulations

a. Summarize the purposes for formulating insecticides.
b. Describe how the common formulations of insecticides are made.
c. Explain the need for various formulations of insecticides.

Insecticide Nomenclature, Classification, and Modes of Action

a. Classify the organochlorine, organophosphate, and carbamate insecticides into their major subgroupings.
b. Describe the modes of action for the organochlorine, organophosphate, carbamate, and formamidine insecticides.
c. Identify insecticide classifications based on the chemical structures of insecticides.

Factors Affecting Insecticide Efficacy

a. List several factors that influence insecticide efficacy.
b. Explain the fallacies of early and light insecticide applications and that of late cleanup applications in controlling subeconomic infestations of crop plants by insects.

Insecticide Application and Drift

a. Select the appropriate formulation for drift control.
b. Predict the drift characteristics of different application methods in a 3-mph breeze.
c. Explain the advantages and disadvantages of ground versus aerial applications of insecticides.
d. Arrange the various formulations in order of increasing off-target drift.

Selective Use of Insecticides

a. Clarify the meaning of selective insecticides.
b. Differentiate between physiological and ecological selectivity.
c. Compare the value of selective versus broad-spectrum insecticides.
d. Describe ways of achieving selective action with insecticides.
e. Discuss methods of improving selectivity of insecticides.

Legal Aspects of Insecticide Use

a. Explain the purposes of the Federal Insecticide, Fungicide, and Rodenticide Act of 1947, the Miller Amendment of 1954, and the 1972 Federal Environmental Pesticide Control Act.
b. Describe the eight basic provisions of the Federal Environmental Pesticide Control Act.
c. Outline the instructions provided on a typical insecticide label.
d. Explain the purpose and meaning of field reentry safety intervals.
e. List the two key sources of information regarding pesticide poisonings and accidents.
f. Compile a list of at least ten safety rules appropriate for your own work situation.

D-Host Plant Resistance

a. Describe host-plant resistance.
b. Explain degrees of plant resistance.
c. Differentiate between host-plant resistance and apparent resistance.
d. Describe how host-plant resistance fits in the insect-pest-management system.
e. List the prominent advantages and limitations of host-plant resistance.

E-Physical and Mechanical Controls

a. Distinguish between physical and mechanical controls and other forms of insect control.
b. Explain why physical and mechanical insect control methods have not been used more extensively.
c. What is the value of a selective control method that is only 50% effective against a target species?
d. Propose the practical uses for ultraviolet light traps in insect pest management.

F-Regulatory Control

a. Discuss the fundamental regulatory control principles.
b. Designate the legal authority for regulatory action at the federal and the state levels.
c. Identify the three types of regulatory control programs.
d. Explain the importance of regulatory control to insect-pest-management programs.

Unit 4 POTENTIAL COMPONENTS OF THE INSECT PEST MANAGEMENT SYSTEM

A-Microbial Control

a. Clarify the place of microbial control in a pest management system.
b. Compare microbial control with biological and chemical methods.
c. Discuss the major limitations and advantages of microbial agents as a means of insect control.

B-Insect Pheromones

a. Identify the practical uses for sex pheromones in insect-pest-management systems.
b. Distinguish between lures and pheromones.
c. Describe some of the potential uses of sex pheromones in insect-pest-management systems.

C-Chemosterilants

a. Explain the advantages of chemosterilant field use over rearing and release of sterilized insects.
b. Summarize the modes of action of the chemosterilants.
c. Compare the hazards of using conventional insecticide control with that of chemosterilants.

D-Insect Growth Regulators

a. Explain why insect growth regulators are highly compatible with insect-pest-management systems.
b. Identify the classes of insect growth regulators showing potential in insect-pest-management systems.
c. Differentiate between first-, second-, and third-generation insecticides.
d. Indicate the primary difficulty in using juvenile hormones for insect control.

Unit 5 THE PRACTICE OF INSECT PEST MANAGEMENT

A-Establishment of Insect Pest Management Programs

a. Discuss the importance of sampling in developing the insect-pest-management system.
b. Identify the personnel involved in insect pest management.
c. Enumerate the duties of the various personnel groups.
d. Explain the importance of each personnel group in effectively developing insect pest management.
e. Prepare a simple plan for establishing an insect-pest-management system.
f. Identify the priorities necessary in developing the insect-pest-management system.

B-Some Existing Insect Pest Management Programs

a. Identify the essential components of these programs.
b. Compare the different programs relative to possible improvements.

PRACTICAL INSECT PEST MANAGEMENT

UNIT 1 THE NATURE OF INSECT PEST MANAGEMENT

What is insect pest management? How does it differ from insect control? What makes it suddenly so important, even popular? How and when did it start? Is it compatible and usable in today's technical agriculture? These are some of the questions whose answers we seek in this unit.

A thorough discussion of agricultural pests and insect pest management must, of necessity, include all organisms that adversely affect man's efforts to produce food, feed, and fiber crops. Among the more important of these are weeds, vertebrate pests, plant pathogens (including nematodes), mites, and insects. Such a treatment of all agricultural pests would be an ambitious undertaking, far beyond the scope of this book. We shall deal only with insect and mite pests.

Of the insect species found in any farm community, only a small percentage are "pests." What, then, is an insect pest? Relative to agronomic and horticultural crops we could define such pests as those species which routinely or occasionally cause damage resulting in reduced yield or quality of the marketable product. Such a definition would exclude most species found in the farm community. Many of these non-pest species play no particular role from the standpoint of man's interests, but many others do, and some have the potential to affect crop-plant production directly. These would include the pollinating insects, the predators and parasites of insect pests, and the "potential" pests, all of which can be affected by man's activities in producing his crops. It will soon become apparent in this manual that it is equally important when managing the pest insects and mites to employ management practices that enhance the survival rate of beneficial insects and mites and also prevent the elevation to "pest" status of those having the potential to become pests.

Before launching into a discussion of insect pest management it seems appropriate to clarify some key terminology. Although these and other terms appear in the Glossary, it is of utmost importance that we proceed through the manual from the common foundation that a mutual understanding of these terms provides:

(a) Pest: Any organism that injures or causes damage to man's interests. In this book the word is used in a more restricted sense to refer only to insects and mites that cause economic damage to agricultural crops.
(b) Insecticide: Any substance used to kill insects.
(c) Agroecosystem: An agroecosystem is an agricultural area sufficiently large to permit long-term interactions of all the living organisms in their non-

living environment.

(d) Insect control: The performance of any practice that prevents further increase in pest population growth or that suppresses or reduces existing pest populations.

(e) Insect Pest Management (IPM): The practical manipulation of insect or mite pest populations using any or all "control" methods in a sound ecological manner.

(f) Integrated control: The integration of the chemical and biological control methods.

(g) Economic level: This term is used to denote the insect or mite pest level at which additional IPM practices must be employed to prevent economic losses, resulting from loss of yield or quality of the marketable crop.

(h) Pest-management specialist: A person capable of managing pests in a farm community. Qualifications of such a person will include a thorough knowledge of the taxonomy, biology, and ecology of the more important species; familiarity with the characteristics of available pest control chemicals; sampling techniques; economic pest levels for the various crops; and a good understanding of the production and growth characteristics of the crops involved.

Insect pest management (IPM) brings together into a workable combination the best parts of all control methods that apply to a given problem created by the activities of pests. A somewhat more scientific definition of IPM would be, "the practical manipulation of pest populations using sound ecological principles to keep pest populations below the level causing economic injury." The emphasis here is on "practical" and "ecological." There are many ways of controlling insect pests, but only a few are practical. And fewer yet are ecologically sound, such that a worse situation is not created. IPM, then, is "putting it all together"—using the best combination of control techniques to allow us to "live" with the pest while incurring no economic losses.

Another term we frequently encounter is "integrated control." It is often used interchangeably with IPM, though in the strictest sense these terms are not identical. The distinction between IPM and integrated control is real, though small, when compared with the old idea of "insect control," which tends to be dependent on a single practice, especially chemical control. Originally, integrated control simply meant modifying chemical control in such ways as to protect the beneficial insects and mites. In other words, integrating chemical and biological control methods. Later this concept was broadened to include all suitable methods that could be used in a complementary way to reduce pest populations and keep them at levels which cause no economic damage. This is IPM.

1.1 How do the terms insect pest management, integrated control, and insect control differ?

- - - - - - - - - - - - - - - - -

"Insect control" to many, denotes chemical control. "Integrated control" combines chemical and biological methods. "Insect pest management" utilizes all available tools and techniques.

Both integrated control and IPM are significant advances in crop protection when compared with "insect pest control," which had come to mean, "treat with insecticide when the insect pest is present in an effort to kill all the insects on the crop." IPM includes a variety of options, any one of which may not significantly reduce the pest population, but the sum total of which will give adequate reduction to prevent economic losses. This does not mean that the pest would be eliminated from the crop or the area. In fact, with IPM it is usually desirable to maintain low levels of the pest at all times. Reasons for this will be explained later.

IPM is not a static, unyielding system. It is dynamic, ever changing as we develop a better understanding of all the factors that affect the system. These factors include climate, alternate host plants, beneficial insects, and man's activities. In a narrow sense, IPM means the management of the few important pests generally found on our crops, but consciously or not, it must include all insect pests—not only the "key" ones but also the secondary pests, which seldom do any harm. If this were not so, we might suddenly find some of these minor insect pests or even nonpests elevated to the status of serious insect pests because of our failure to consider them in the total scheme.

Examples to be presented later illustrate the advantages of using combinations of several management practices to suppress pest populations. This does not necessarily mean that the practices all function at the same time; different ones may be appropriate at different times of the year. All, however, contribute to the total effective suppression of the pest, even though the effect of any one practice may be relatively small. By using such combinations the problems associated with dependence upon a single practice can be avoided.

1.2 IPM is an ____________________ system, and all management practices (do) (do not) have to function at the same time.

- - - - - - - - - - - - - - - - -

- - - - - - - - - - - - - - - - - -

ever changing do not

IPM as a concept is not new. Only the name is. Many of the components of a sound IPM system were known some fifty years ago through the research of Dwight Isely in Arkansas. His extensive and foresighted work with cotton insects and mites in the mid-1920's was sufficient to provide a sound basis for today's IPM. His management of such pests as the boll weevil, the bollworm, and spider mites, was based on the principles of applied ecology, the vital foundation of IPM.

1.3 How can you say that pest management is not a totally new concept?

- - - - - - - - - - - - - - - - - -

Because many of the components of a sound IPM program were being used 50 years ago to effectively suppress insect pests.

After the synthetic organic insecticides became generally available in the late 1940's, those concerned with insect control dismissed completely the need for any method except chemical for controlling insects. It took several years and much wrestling with obvious problems associated with chemical control for it to become apparent that changes were needed.

Thus, in handling insect pest problems, we have gone full cycle from the early, applied-ecology days, to chemical control, to integrated control, and finally to a multicomponent IPM system founded on ecological principles.

1-A THE NEED FOR INSECT PEST MANAGEMENT

Why is IPM needed? Why not continue to control insects as they have been in the past? The answers to these questions are somewhat complicated and yet they must be dealt with and understood before proceeding. First, let's take a look at the recent history of insect control before examining the reasons for IPM.

Introduction of the organochlorine insecticide DDT began an era of insecticidal control of insects. As a result, a vast chemical industry began to grow and prosper. Entomological research and extension work largely emphasized the use of insecticides to control insects. One new insecticide followed another, and new groups such as organophosphates and carbamates appeared. Insecticidal control provided a quick, inexpensive, and convenient method of controlling insects. It greatly slowed or stopped efforts such as Isely's to develop methods of insect control that were the forerunners to the methods used in IPM systems today.

There were many reasons for the need to resume emphasis on the development of IPM. These were brought to light by problems resulting from large-scale use of insecticides.

One of the first problems was the development of resistance or tolerance by certain insects to insecticides used against them. Beginning with the resistance of houseflies to DDT, this problem has continued to increase, and today about 250 species have shown resistance to certain insecticides and some species are resistant to more than one group of insecticides.

After a few years of such use of insecticides, there arose the problem of residues remaining in food and feed crops, in the soil, and in animals. Some insecticides, such as the organochlorines, are highly persistent because of their chemical stability. Others, such as the organophosphates, are much less persistent and are rapidly degraded into harmless compounds. Basically, the differences in persistence and the resulting residue problems are related to the structure of each chemical, although other factors are involved. In order to cope with the residue problem, growers found it necessary to use the more toxic but non-persistent compounds. Although this relieved the residue situation somewhat, it created other problems. The high mammalian toxicity of some of the non-persistent insecticides creates a substantial health hazard to persons handling and applying them. Their broad-spectrum of activity in terms of destroying insects in the target area of treatment and their requirement of more frequent application to maintain insect control resulted in a disturbance of the relationship of beneficial insects to pests, permitting pests of minor importance to rise to major pest status. They also resulted in increased costs since the less persistent insecticides are generally higher priced and more applications are required. These factors have contributed to the need for developing pest-management systems that emphasize alternative methods of control and that also minimize the use of insecticides.

Owing to either insect resistance or residue problems, certain insecticides began to be eliminated, resulting in greater demand for the remaining insecticides. This situation, along with a continually increasing insect problem in certain crops (particularly cotton and lettuce in some areas) brought about a shortage of many of the commonly used and effective insecticides. In addition, the shortage, combined with other inflationary pressures, has caused sharp increases in the price of most insecticides, thus contributing to sharply rising grower production costs.

Another important contributing factor to the shortage and the price increase has been the petroleum situation. Petroleum solvents are a necessary part of emulsifiable concentrate formulations of insecticides. The oil shortage caused the need to conserve available insecticides as much as possible and to use them only when absolutely needed. IPM, with its basic reliance on field sampling and use of economic levels was the best means of reducing use of available insecticides to the minimum required for economic insect control.

1.4 What are some of the major reasons for the need to emphasize the development of insect pest management?

- - - - - - - - - - - - - - - - - -

(1) resistance, (2) residues, (3) hazard, (4) disturbance of relationship of beneficial insects to pests.

1-B PHILOSOPHY OF INSECT PEST MANAGEMENT

In order to develop and use an IPM program, you must first understand the real meaning behind this concept. It involves more than just knowing that a pest is present on a crop and that such an insect should be controlled. You must also find answers that explain why an insect pest normally occurs at certain population levels each year. This requires a thorough understanding of the role of all the factors responsible for a pest population reaching these levels at particular times of the year.

An insect pest population must be analyzed throughout an entire farm community, the agroecosystem, relative to its distribution and expected seasonal population changes as influenced by climatic conditions. All crops in the community should be considered and their role in the seasonal buildup of the pest determined. Also, the level of damage that each crop can tolerate without economic losses should be established. Once these limits have been determined, the next step is to seek ways to keep the insects from exceeding economic levels. Several methods may be used to accomplish this. The first consideration, however, should be given to the methods that provide a permanent adverse effect on the insect pest over the entire farming community. This would include such techniques as host-free periods and other cultural practices (defined in Unit 3). Insect suppression on an area-wide basis would then give growers more flexibility in dealing with their individual insect pest problems.

The ultimate goal therefore, is to reduce the pest status of insects through management of populations. This means that a number of different insects, both good and bad, will probably be present in crops most of the time, but at low levels—thus the "dirty field". Obviously, this method of handling your pest problems is out-of-step with the traditional desire for a pest-free field—the "clean field". The "dirty field" is necessary, however, if IPM is to be based on sound ecological principles.

1.5 The ultimate goal of IPM is to ______________________________.

- - - - - - - - - - - - - - - - -

reduce the pest status of insects through management of populations

---

All agricultural areas are artificial since man is continually changing them. It stands to reason, then, that no natural balance among insects exists as it does in undisturbed areas. There are, however, general groups of insects that are usually found among crop plants and that have little adverse effect on the crops. Conditions may change rapidly in such an agroecosystem, however, if natural control is disturbed.

One of the major disruptive factors in the agroecosystem for the past thirty years has been the use of chemical insecticides for insect control. There are several reasons for this heavy reliance on insecticides, but the detrimental side effects of this single practice have been of such magnitude that they can no longer go unheeded. Changes are called for if crop protection is to continue at desired levels. Many crops require scheduled insecticide applications for protection from certain insects. This simply means that these insect species are getting "out of control".

The first step in developing IPM to handle this out-of-control situation is to devise ways of lowering insect pest populations to levels the grower can tolerate, that is, which will produce high yields of quality crops. This will usually require the use of insecticides, but their use will be more precise and will complement other control tactics integrated into an IPM scheme.

1.6 The first step in developing IPM is to ______________________________.

- - - - - - - - - - - - - - - - -

lower insect pest populations.

The second step, once insect populations have been lowered to managable levels, is to maintain these levels and minimize their fluctuations so that economic levels (= economic thresholds) are never exceeded. This requires a continuous updating of information on the influences of a changing ecosystem on pest and beneficial insect populations. Since any agroecosystem is constantly being altered by man, it may be changed to the pest's benefit by providing it with more favorable host plants or a better sequence of host plants, or to its detriment by adopting practices or cropping sequences harmful to the beneficial biotic complex. This illustrates the importance of a thorough understanding of the biology and ecology of all important species, both good and bad, in the system. With enough information, these changes can be anticipated and compensated for by using different management practices.

1.7 The second step in developing IPM is to ____________ ______________________________________.

- - - - - - - - - - - - - - - -

maintain pests below economic levels

Accomplishing the above steps requires the services of a special kind of person possessing unique qualifications. Does this mean that IPM can only be useful when practiced by skillful scientists, broadly trained in all areas of crop protection? Certainly not, but neither does it mean that a person capable of identifying an insect, and of showing some arbitrary insect density through sampling, possesses the knowledge and expertise to manage insect pest populations. The answer lies somewhere between these two extremes. An individual with a general agricultural background and an intense desire to understand the dynamic nature of an agroecosystem can be trained in the areas of insecticides and insect biology and ecology to fill the supervisory role in IPM. This person would become increasingly knowledgeable with the observations and information gained through day-to-day contact in the field.

We have stressed the complexity of the agroecosystem, some of the basic factors that must be determined, the major objectives of the IPM system, and the unique qualifications of the pest-management specialist responsible for implementing these objectives. Now let us explore the basic elements upon which a sound IPM system rests. These basic elements are natural control, sampling, economic levels, and insect biology and ecology.

1.8 What are the four basic elements of IPM?

- - - - - - - - - - - - - - - -

natural control, sampling, economic levels, and insect biology-ecology

---

The first element of IPM relates to the fullest utilization of naturally occurring suppressive measures, including any practice by man which will make the total ecosystem less favorable for growth of the insect pest population. Obviously, this requires a thorough understanding of the ecosystem.

The naturally occurring suppressive factors referred to in the first element may act directly or indirectly on pest populations. Indirectly, the ecosystem may be managed or altered in such ways as to make the environment more harmful to the pest and thus limit growth of its population. More directly, protection and the use of beneficial insects may help keep potentially damaging insect pest populations at subeconomic levels.

The second element is that of using sound economic levels as the basis for applying chemical control measures. Establishing and using dynamic economic levels provides a basis for delaying the use of insecticides. This permits the maximum utilization of other control methods, such as the use of beneficial insects.

The use of economic levels implies adequate sampling, another basic element, of all harmful and beneficial insects in the agroecosystem and particularly in any one crop at a specific time. The levels found through sampling must then be measured against the economic level established for the crop, the beneficial insects, and the probable population trend of the pest species. The sampler thus becomes a key person in an IPM system.

The fourth element, insect biology and ecology, is essential to the fullest utilization of the other three elements. Little concerning natural control can be understood without detailed knowledge of the biology and ecology of all the species present. This knowledge is also essential in establishing the role of each species in the system and in determining the amount of damage the pest species are doing. Adequate sampling is directly dependent upon a thorough familiarity of the species involved.

One of the obstacles in making an IPM program work is lack of understanding and acceptance of its concepts by the grower. For years he has been schooled in the one-hundred-percent-kill concept. The old adage, "The only good bug is a dead bug" has prevailed too long. Consequently, it has not been easy to convince him to accept the dirty-field technique, something less than complete kill. Growers who have been involved in IPM programs for a number of years have gained an understanding of and confidence in the system and now rely on insecticides as only a part of their total insect-control program.

Another obstacle that has handicapped the IPM approach has been the lack of knowledge of adequate economic levels. And, too, many growers or their representatives have failed to utilize the available economic levels. This is somewhat understandable in some instances because, at best, the economic levels have been poorly defined.

Regardless of how poorly defined they were, however, most or all were set to prevent any chance of loss and were thus more desirable than scheduled insecticide applications for the entire growing period of the crop. Lack of economic levels has resulted in much overtreatment. The philosophy, "When in doubt, treat" has brought on many of the problems attributed to insecticides.

1.9 Name two important obstacles to IPM.

- - - - - - - - - - - - - - - - - -

lack of grower understanding and acceptance and lack of knowledge concerning economic levels

---

How do we overcome these obstacles and use the elements discussed above to establish a workable IPM program? We must fulfill the interrelated requirements of all the components in the system. Certain of these are necessary for the sound functioning of an IPM program. These requirements have been discussed to a certain extent in earlier parts of this unit, but are given here to show their interdependence. They are 1) knowledge of most insects and crops in the system, 2) qualified personnel for running the program, and 3) practical procedures for applying management practices. A minimum of scientific knowledge required would include 1) the general biology, distribution, and behavior of the key insect pests, and why they are pests; 2) an estimate of the insect pest population levels that can be tolerated without significant crop loss; 3) a rough evaluation of the benefits obtained from the major insect predators, parasites, and pathogens; and 4) information on how the use of insecticides and other control measures will affect both the insect pests and their natural enemies.

The more IPM specialists know about the entire agricultural area and the organisms involved, the better equipped they are for making decisions themselves or supplying the appropriate information to the individuals who do make the decisions. For example, relative to economic levels, they should know the relationship between insect numbers and crop development, and they should certainly be aware of the best sampling techniques for population assessment. They should also be aware of the alternatives available to help handle the different pest problems (for example, techniques to take full advantage of biological control agents and cultural control practices), and they should be familiar enough with insecticides and their manner of application to permit flexibility in their use to achieve selective action.

1-C HISTORICAL EXAMPLES OF SINGLE-COMPONENT INSECT PEST CONTROL

IPM, as you now understand it, is a multicomponent approach, usually involving several control strategies or techniques. There are, however, historical milestones of successful insect control involving only a single technique, now considered as only one tactic in the much broadened IPM concept. Several classic examples of single-component control are presented below. These prove both interesting and useful in exploring the history and development of IPM.

1.10 How are single-component control techniques useful?

- - - - - - - - - - - - - - - -

As a part of the IPM system.

One method that excites the imagination is the field of *biological control*—the use of predators, parasites, and pathogens to suppress insect pest populations. The classical example that established biological control as a valid method of pest control was the use of vedalia beetles to combat cottony-cushion scale. In 1888, the vedalia ladybird beetle[1] was introduced into California from Australia, the native home of the cottony-cushion scale, in an attempt to control this pest, which had virtually ruined the citrus industry. Within months, the predator spread throughout citrus groves and completely controlled the cottony-cushion scale.

Developing an area-wide *planting date* of a single crop has served to avoid host-specific insect pests. Before 1900, the Hessian fly was a devastating pest of wheat. After the biology of the fly was thoroughly investigated, it was found that the planting of wheat in large areas after all first-generation flies had died would virtually eliminate infestations and yet permit the normal production of a wheat crop. This became the classic "fly-free planting date," an extremely useful cultural technique, later adapted for the control of several crop pests, including the sorghum midge on sorghum.

Insect-resistant crop varieties are now common in many cultivars. However, as late as the mid-1950's, the European corn borer successfully infested corn crops in the North Central states, causing the stalks to weaken and break, making harvest difficult, and resulting in heavy yield losses. With the development of resistant

---

[1]Only insect common names, where available, will be used in the text; a list of common and scientific names appears in the Appendix.

varieties of corn that could withstand the first generation corn borer and remain upright, corn yields were increased measurably, by as much as one-third in some areas, merely by permitting the harvest of ears formerly lost to lodging.

The use of insect *disease pathogens* is increasingly becoming a useful method of crop protection. The first successful propagation of an insect disease pathogen was developed in 1934. The causal organism of the Japanese-beetle milky disease was identified as a spore-forming bacterium. Shortly afterward, the organism was mass-produced and distributed for application to soil where the beetle larvae became infected and quickly died.

In 1955, the first successful eradication of an insect by the *release of sterile males* was accomplished on the Caribbean Island of Curacao. The effort involved the releasing of sterile male screwworms into a normal population so that females with which they mated would lay predominantly sterile eggs. This small-scale eradication of the screwworm was expanded to the southern and southwestern United States to suppress this pest of cattle. The program now also includes all of Mexico.

The practice of destroying insect pests while they are concentrated in relatively small areas has been used effectively. *Trap cropping* has provided one such means whereby the pest could be killed with limited use of insecticide or could be destroyed along with the small acreage of the trap crop. Early trap cropping of boll weevils was demonstrated in Arkansas by Isely in 1934 when an early-maturing cotton variety was planted in small plots within cotton fields. The overwintering weevils congregated on these early plants which were then treated three or four times with calcium arsenate dust. This would often prevent the damaging populations from developing within the surrounding fields. It was also shown that boll weevils in hibernation could be reduced by burning or clearing around cotton fields, since they usually hibernate near their host plants.

Crop rotation is effective against some insects that spend their immature stages feeding on roots. Corn rootworms are some of the more destructive pests in the corn belt and are readily controlled by crop rotation. Most of the eggs hatch in the spring in fields planted with corn the previous year. If a different crop is grown in these fields, the young rootworms die because of a lack of suitable food plants.

The list of examples goes on and on and is impressive, particularly as we retreat in time. (Our forefathers' ingenuity was astounding at a time when chemical control measures were unavailable or available only on a limited scale.) However, as we acquire more information concerning insect ecology and behavior, systems of pest management based on the single-factor approach can be enhanced by supplementary inputs. Thus, the playing field of the future will actively engage entomologists in a multicomponent strategy—insect pest management.

1.11 In two or three words each, list five of the single-component insect-pest controls used as examples.

- - - - - - - - - - - - - - - - -

Biological control, planting date, resistant varieties, disease pathogens, sterile male release, trap cropping, and crop rotation.

References

Anonymous. 1969. Insect-pest management and control. Nat. Acad. Sci. Publ. 1695.

Isely, D. 1924. The boll weevil problem in Arkansas. Ark. Agr. Exp. Sta. Bull. 190.

Isely, D. 1943. Relationship between early varieties of cotton and boll weevil injury. J. Econ. Entomol. 27: 762-766.

Newsom, L. D. 1974. Pest management: History, current status and future progress. In Proceedings of the Summer Institute on Biological Control of Plant Insects and Diseases, ed. F. G. Maxwell and F. A. Harris. University Press of Mississippi State University, Starkville, Miss.

Rabb, R. L. and F. E. Guthrie, eds. 1970. Concept of pest management. Proceedings of the conference held at North Carolina State University, Raleigh, N.C., Mar. 25-27.

Smith, R. F. and H. T. Reynolds. 1966. Principles, definitions and scope of integrated pest control. Proceedings of the FAO Symposium on Integrated Pest Control 1:11-17. Rome, Oct. 1965.

Stern, V. M., R. F. Smith, R. van den Bosch, and K. S. Hagen. 1959. The integration of chemical and biological control of the spotted alfalfa aphid: The integrated control concept. Hilgardia 29(2):81-101.

Whitcomb, W. H. 1970. History of integrated control as practiced in the cotton fields of the South Central United States. Proceedings of the Tall Timbers Conference on Ecological Animal Control by Habitat Management, No. 2:147-155, Feb. 26-28.

UNIT 2 BASIC ELEMENTS OF INSECT PEST MANAGEMENT

The four elements basic to any IPM program are *natural control, sampling, economic levels*, and a detailed knowledge of the *biology and ecology* of all important insects in the system. Each element is vital and provides the supporting role for all components that can be fitted into any insect-pest-management system (Fig. 2.1). Let us now explore in detail each of these elements according to what they are and how they are important in an IPM program.

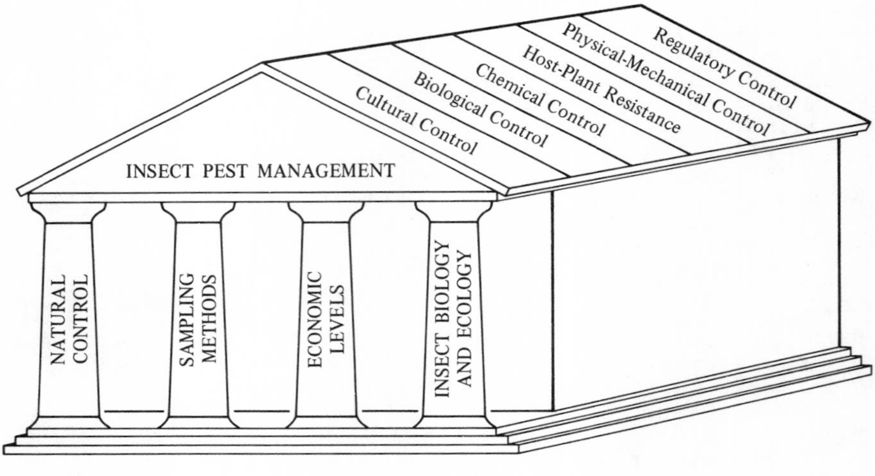

Figure 2.1 Diagrammatic illustration showing the supporting nature of the four basic elements to insect pest management and its six components.

2-A NATURAL CONTROL

The meaning of natural control may vary among individuals, depending upon their experience and background. To some, it means control by predators and parasites, or naturally-occurring biological control. This, of course, is only a part of natural control. To others, it simply means that the insect occurs at such low numbers that it is not a pest. In the latter case, the reasons why the insect is not a pest receive little thought—it just happens that way. Since neither of these explanations is adequate, a deeper understanding of the complex system and the factors influencing that system must be achieved. Natural control as used in this text refers to the long-term suppression of a population resulting from the actions of the total environment.

2.1 How does biological control relate to natural control?

- - - - - - - - - - - - - - - - - -

Naturally-occurring biological control is only a part of natural control.

The broad limits of the existence of natural populations, i.e., natural control, have been set in a number of different definitions or concepts. All, in a very general sense, indicate that a natural population will not continue to grow indefinitely nor will it decrease to extinction. A definition such as this tells what happens but not how it comes about. A more desirable definition would specify general population bounds (that is, the limits above or below which the population would not normally occur) and would tell how they happen. A comprehensive definition would indicate that bounds come about by the combined actions of the whole environment, including both the physical and biological components.

2.2 What is natural control?

- - - - - - - - - - - - - - - - - -

Long-term supression of a population by the total environment.

Part of the difficulty arising in an explanation of natural control is the terminology involved. Key terms generally used to describe an animal population are balance, balance of nature, and equilibrium positions, which may have entirely different meanings to different individuals. The terms are useful in establishing a general relationship of the numbers of an insect species to its general environment and to other species. For example, the normal levels of pea aphid and the corn earworm in alfalfa are quite different, but the general levels that each may reach is characteristic for the respective species.

Balance, according to some, simply means that populations vary, increasing during favorable conditions and decreasing during unfavorable ones. The climatic and soil conditions place natural limitations on insect growth, and the population, in turn, is biologically limited in capacity to exist in these varied conditions. According to this idea, in the long run population increases and decreases balance out but this is not a very valid explanation of natural control.

The relative success of any insect species is dependent upon many factors. First, a species is adapted to certain physical conditions, and the environment that sets the stage for that species directly affects population growth rate, including longevity, oviposition, mating, dispersal, and other population characteristics. Second, as the population adjusts to the physical environment in terms of its growth potential, other factors are acting on the population in a suppressive manner. These are mostly biological factors, such as predators and parasites, and are related to the density of the population.

2.3 What are the two forces of the environment involved in natural control?

- - - - - - - - - - - - - - - -

physical conditions, for example, temperature, wind, soil type

biological factors, for example, food plants, predators, parasites

---

This view of balance of populations is one that includes some kind of buffering system. This simply means that when some condition causes a population to increase or decrease from its general long-term level, there is a tendency to return to the original equilibrium position. Even in a fairly constant physical environment, populations will fluctuate about a general level, and this is balance resulting from the action of specific mechanisms. In mixed-cropping communities, there is greater stability and thus greater balance than in simple pure-stand communities. This results from a greater number of checks and balances.

In IPM we are often concerned with a single insect pest species. Therefore, we are interested in *balance* of that species. So, we are usually concerned with the mechanisms of natural control of single-species populations. This, of course, is affected by the overall balance of a community of which a given population is a pest.

In any given cropping system, certain species are fairly abundant, others occur more sporadically, and still others are rarely encountered. This condition, which may occur within certain fairly predictable limits year after year unless some major change occurs in the environment, describes a status of variation of a species. The absolute numbers may vary from time to time, but the numerical relationship between the several species remains more or less stable. The process by which this is accomplished is *natural control* and includes the action of the whole environment, including density regulating elements in relation to the conditions of the environment.

We have described the two main views of natural control. The first and most commonly accepted says that as a population increases in numbers it induces repressive forces which tend to prevent further increase. These repressive forces can come from the environment or from the population itself. The counterpart to this view is that as the population density declines, there is less stress from the repressive forces and the population tends to recover and start to increase again.

The second view of natural control is primarily that of chance existence. This means that periods favorable for population growth are counterbalanced by periods of unfavorable conditions whereby populations decline. Populations thus rise and fall accordingly, and, in the long run, balance out. This view, then, implies a state of 'balance" in the environment, but offers no scientific explanation as to how it comes about.

In the first view, balance is seen as occurring in the population, with an appealing and logical explanation of how it can happen. This is our view of natural control as it happens in an agroecosystem.

The main thought concerning natural control is that there are two separate forces or sets of factors operating together. One of these involves the physical environment, which sets the environmental load limit (determination) and is not influenced by the numbers of the organisms. The other involves stresses related to the numbers of the organisms. One situation may be almost continuously optimum for growth of a species, and the species will occur at much higher levels than in a harsher environment. Regardless of the type of environment, regulation of an insect about its general equilibrium level, whether it be high or low, comes about only by the action of density-dependent factors (for example, predators or parasites). In this case, as the population increases, an increasingly larger percentage will be killed by predators or parasites, or perhaps greater resistance to population growth will result from reduction of the food supply or from some detrimental effects caused by the population itself.

2.4 What is the difference between population determination and regulation?

\- - - - - - - - - - - - - - - - -

> The physical environment determines or sets the load limit or carrying capacity of an environment. Regulation occurs as a result of the action of biological agents; the two operate together.

General population levels are set (determined) and maintained (regulated) in different types of environments by physical and biological factors, respectively. Year in and year out, variations in climatic conditions determine differing population levels of many pests. In many cases, however, biological factors such as predators and parasites operate to regulate populations regardless of the general level set by the environment. In some years insect pests exceed economic levels, and additional means have to be employed to keep the pests within economic bounds. Regardless of the initial level set by environmental conditions, if all regulatory agents are removed, such as destruction of beneficial insects with insecticides, the general environmental conditions are permissive for new, higher pest levels. This type of upset occurs routinely in agricultural systems where insecticide applications are made.

Whether you agree with these or other theories of natural control, logic dictates the acceptance of some explanation for the fact that population growth does not continue indefinitely. Man's activities may change an agroecosystem to provide for greater or lesser environmental capacity for insect pest numbers. It is this capability that man has for altering the insect's environment to make it less suitable for insect pests that offers much hope for IPM.

References

Doutt, R. L. and P. DeBach. 1964. Some biological control concepts and questions. Pp. 118-142 in *Biological control in insect pests and weeds*, ed. P. DeBach. Reinhold Publishing Corporation, New York.

Huffaker, C. B. and P. S. Messenger. 1964. The concept and significance of natural control. Pp. 74-117, ibid.

Nicholson, A. J. 1954. An outline of the dynamics of animal populations. *Australian J. Zool*. 2:9-65.

Solomon, M. E. 1949. The natural control of animal populations. *J. Anim. Ecol*. 18:1-35.

Thompson, W. R. 1929. On natural control. Parasit. 20:90-112.

Thompson, W. R. 1956. The fundamental theory of natural and biological control. *Ann. Rev. Ent*. 1:379-402.

Uvarov, B. P. 1931. Insects and climate. *Trans. Ent. Soc. London* 79:1-247.

2-B SAMPLING

The development of good, unbiased sampling is a prerequisite to rational insect control and especially to a full utilization of IPM. A means of determining the approximate numbers of insects is absolutely necessary to provide information upon which management decisions can be made. Sampling and economic levels work together; one is of little value without the other. It is necessary to know the insect population levels and the economic levels to make meaningful IPM decisions.

The requirements relative to the methods and techniques of sampling used for IPM purposes may be somewhat different than the more elaborate and statistically sound procedures used for research. Factors such as time requirements and economic considerations make it necessary to establish the most practical sampling program possible for each crop and situation involved. Usually IPM programs utilize the simplest sampling methods that provide the information needed for insect control decisions. Although the importance of improved sampling is recognized, the following discussion of sampling gives primary consideration to methods presently used in IPM.

2.5 Why is sampling necessary in IPM?

- - - - - - - - - - - - - - - - -

Sampling provides a means of determining approximate insect numbers for making IPM decisions.

Types of Sampling

There are several ways of sampling crops for insect numbers or damage. These include random, sequential, point, and trap sampling. Any of these can be effectively used, depending on the crop and pest situation involved. The random sample is the most commonly used method of determining insect populations. As the name implies, the sample is taken at random, with good field coverage, to determine insect numbers or damage per sample unit. This is often recorded as percent infestation or damage without consideration of the total number per acre. Examples of this type of sampling are (1) counting the number of bollworm larvae per one hundred randomly selected cotton terminals and (2) counting the number of lygus bugs caught by one hundred net sweeps in seed alfalfa.

2.6 Four ways of sampling crops for insect numbers or their damage are ____________________, ________________, ____________________, and _______________sampling.

- - - - - - - - - - - - - - - - - -

_ _ _ _ _ _ _ _ _ _ _ _ _ _ _ _ _ _ _ _

random, sequential, point, and trap

Point-sample is a measure of population density designed to relate the number of insects or their damage to the number of plants or plant parts per acre. This has permitted establishing economic levels based on numbers of insects or insect damage per unit of measure as related to stage of fruiting or plant development. The following procedure is suggested to scout for boll weevils by the point-sample method: Examine the first fifty squares (fruiting buds) 1/4 inch or larger in diameter, and measure the row feet required for the sample. Record the number of squares punctured or damaged by boll weevils, and repeat the procedure in at least three other locations in average-sized fields to obtain representative coverage. The number of damaged squares per foot of row is then related to the total squares converted to the number per acre.

Sequential sampling is different from other methods in that it requires continued sampling until a preestablished upper or lower infestation level is found. For example, if an insect species is to be controlled when there are five per fifty plants or not controlled when there are three per fifty plants, sampling is continued as long as the count remained between three and five. Sampling is stopped when either of these limits is detected or when a preestablished number of plants have been sampled, even when the count indicates continued sampling. This method of sampling is especially effective when population levels are either high or low but is not widely used in IPM.

Trap sampling—using light, suction, sticky materials, or sex pheromones—has been used most in the past to detect the presence of insects within an area. Recently, however, sex pheromone traps have been designed and developed for determining field populations as a means of making insect management decisions. Sex pheromones are chemical substances emitted by the insect to attract a mate and are discussed in detail in Unit 4. These chemicals are identified by scientists and produced synthetically for use in baiting traps designed to catch the attracted insects. Gossyplure is used to attract and capture male pink bollworm moths in cotton. Traps are placed in cotton at the rate of one per 20 acres and the number of moths captured is counted and recorded daily. A similar procedure is used to trap codling moth males in apple orchards. Light, suction and sticky traps are sometimes used to monitor flying insects or to collect large numbers of insects. Additional development of techniques will be necessary before the trap method of sampling is generally used in IPM programs.

2.7 __________ ____________ traps are being developed as a means of determining insect populations for making control decisions.

\- - - - - - - - - - - - - - - -

Sex pheromone

Collecting and Observational Methods

Determining a population may be accomplished by counting the number of insects present, or by making damage counts and relating them to the population. Several methods including the sweep net, the drop sheet, and the visual count, are used in measuring or counting the insects present.

2.8 How are insect populations estimated without actually counting their numbers?

\- - - - - - - - - - - - - - - -

By damage counts.

The sweep net is commonly used in catching highly mobile insects such as the plant bugs. The standard sweep net is 15 inches in diameter and its use is determined by the crop involved. For example, in alfalfa and other solid plantings a sweep is considered a 180° arc. In row crops such as cotton, a sweep is one pass of the net across the plants on a single row. When plant height permits, a sweep should be made so that about 10 inches of the plants extend above the bottom of the net. A limitation to the sweep net method is the difficulty involved in standardizing the sweep.

2.9 The standard sweep net is _______ inches in diameter and the standard sweep in solid-planted crops is a __________ arc.

\- - - - - - - - - - - - - - - -

15 180°

The drop sheet can be used to sample insects that dislodge easily from their host plants or that have no wings or lack the tendency to fly when disturbed. Some plant bugs and other insects are sampled in cotton by placing a white sheet of muslin or similar material between two rows and vigorously shaking the plants from 18-inch portions of both rows over the sheet. The insects that drop onto the sheet are counted and the procedure is repeated to give a total of ten counts in a field. The insect count multiplied by five hundred gives the number on a per-acre basis when rows are spaced 40 inches apart. This method has proved particularly useful for counting worms in soybeans.

The visual count method is the most commonly used procedure for determining insect populations or their damage. It involves the counting of insects or their damage by examining a specified number of plants or plant parts. For instance, the spotted alfalfa aphid is visually sampled by counting the number of aphids on three trifoliolate leaves, each selected at random from the top, middle, and bottom of alfalfa plants. The sample should be replicated ten times at random intervals of twenty to fifty walking steps in a diagonal pattern across the field. Bollworms in cotton are sampled by counting the number of larvae and eggs in the terminals of 14 feet of row or the number in twenty-five randomly selected terminals. Either method used is repeated four times in average-sized fields to provide good field coverage. When the 14-row-feet method is used, the results can be multiplied by 250 to obtain the number per acre. (The random sample, on the other hand, gives a percentage of infested terminals.) There are many variations to the visual count method, but fortunately the procedures are usually standard within an area. Standardization is less of a problem with this method than with the sweep net method.

2.10 The ____________ ____________ method is the most commonly used procedure for determining insect populations.

- - - - - - - - - - - - - - - - - -

visual count

Stage of Insect Development and Sampling

Most sampling for IPM purposes is directed at stages of insect development that cause damage to the crop involved. Sometimes, however, counts of eggs and adults are made simply to serve as indicators of expected insect populations. For example, cotton bollworm eggs are normally counted to give information on whether bollworm numbers can be expected to increase or decline in the next several days. Supervisors may even want to know whether the eggs are predominantly white or dark in color, since dark eggs can be expected to hatch in hours, whereas the newer white eggs will require two to three days to hatch. Trapping and counting codling moth adults provides information on generation peaks and permits timely control measures to be taken against the destructive larval stage.

2.11 Why are counts sometimes made of insect eggs?

\- - - - - - - - - - - - - - -

Counts serve as indicators of expected insect populations.

It is usually preferable to sample for the destructive stage of an insect or its damage, since predictions based on numbers of an earlier stage are not always reliable. Climatic conditions, beneficial insect numbers, and other factors can greatly affect the development of an insect population, sometimes changing a potentially serious infestation into a harmless one.

2.12 It is usually preferable to sample for the ______________ stage of the insect or its damage.

\- - - - - - - - - - - - - - -

destructive

The immature stages of insects, known as larvae or nymphs, are usually the destructive stages of pests and are therefore sampled. In many cases, such as aphids and lygus bugs, both the young and the adults cause damage; thus both stages are counted.

Sampling for pupae is not common in population measurement. Soil or crop residue samples are sometimes taken to help estimate the abundance or expected infestation of an insect. Counts of European corn borer larvae and pupae in corn stalks, for example, are used to provide information for estimating their damage.

Sample Size

The amount of sampling necessary depends upon the crop and its pests. Almost invariably, uniformity of infestation does not occur; thus it is essential to take a representative sample that largely overcomes the lack of uniformity. The standard procedure for average-sized fields of crops such as cotton is to sample four areas avoiding sides and ends, but noting insect activity at any point (Fig. 2.2). In random sampling, counts of twenty-five in each area are common to give a convenient total of one hundred for the entire field. Fields larger than 80 acres are sampled in at least two additional areas; those smaller than 20 acres are often sampled in only three or even only two areas. Some crops such as alfalfa and certain vegetables are sampled as the sampler moves diagonally across the field. The most important consideration in sample size is good field representation or coverage.

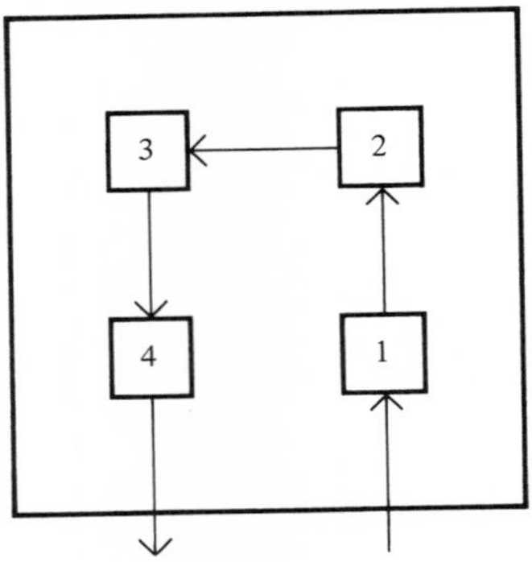

Figure 2.2 Diagrammatic example of procedure for sampling an average-sized field.

2.13 __________ areas are usually sampled in an average-sized field.

\- - - - - - - - - - - - - - - - -

Four

Sampling Records

If data obtained in sampling are to be effectively used, they must be recorded by the scout on a field record form (Fig. 2.3). The form should be designed to provide the needed information on both harmful and beneficial insects, identification of the field, date sampled, and pertinent comments. The form may also be designed to include data on plants and plant parts, but this information may be recorded on a separate form. Since the completed record form is important to the grower and IPM supervisor in making control decisions, it should present data clearly and should include only essential information. Duplicate copies are usually made, one for the scout, one for his supervisor. A well-designed record form is a primary prerequisite in development of an IPM system.

COTTON

COOPERATIVE PEST MANAGEMENT PROGRAM

OWNER ____________

DATE 4-8 ____________

SAMPLER 1-3 ____________

											SUGGESTED TREATMENT LEVEL
	Grower Field Designation										
	Program Field Number		12-15								
	Time: To nearest hour		10-11								
Number per 100 plants examined	Pink Bollworm	Damaged Bolls	16-17								15 infested bolls/100
		Larvae Small	18-19								
		Larvae Large	20-21								
	Bollworm	Larvae	22-23								10-12 larvae/100 plants
		Eggs	24-25								
		Damaged Sq.	26-27								
	Leaf Perforator	Leaves	28-29								25 infested leaves/100
	Cabbage Looper	Leaves	30-31								
	Other										
Number per 100 net sweeps	Lygus	Damaged Sq.	32-33								25 damaged squares/100 or 15-20 lygus/100 sweeps
		Adults	34-35								
		Nymphs	36-37								
	Other										COMMENTS:
	BENEFICIAL										
	Lady Beetle	Adult	38-39								
	Collops	Adult	40-41								
	Orius	Immature	42-43								
		Adult	44-45								
	Big Eyed Bug	Immature	46-47								
		Adult	48-49								
	Nabid	Immature	50-51								
		Adult	52-53								
	Lacewing	Immature	54-55								
		Adult	56-57								

Figure 2.3. Example of a Form for Recording Vital Information When Sampling Cotton

---

2.14 The primary information provided on a field record form includes ____ ____.

- - - - - - - - - - - - - - - - -

counts of harmful and beneficial insects, identification of the field, the date sampled, and pertinent comments

---

Sampling Time and Frequency

Routine sampling for most insects is done at weekly intervals, with more frequent samples being taken when a pest population approaches a borderline stage with respect to economic level. Insects that develop economic infestations quickly under certain conditions, such as the cotton bollworm, must be sampled twice weekly or more often when there is a threat of infestation increase. Sampling at shorter intervals is also required for some crops, such as vegetables, where the economic level of a pest may be very low. Sex-pheromone trap catches are sometimes recorded daily in order to pinpoint increases in moth activity.

2.15 When is it necessary to sample oftener than weekly?

- - - - - - - - - - - - - - - - -

When the infestation is borderline with respect to economic level, and in crops where the economic level is very low.

The time required to sample a field also depends on the crop and its pests. Less time is required in cotton, for example, when the plants are small than during mid- or late-season, and less time is required in areas not infested with boll weevils or pink bollworms than in areas where these insects are present. Where visual counts of insects or their damage in a crop are necessary, a time period of thirty to forty-five minutes by a single scout is usually necessary for each field sampled.

2.16 The time required for sampling depends on the __________ and the ____________ involved.

- - - - - - - - - - - - - - - - -

crop insects

Sampling Personnel

Field samplers are commonly referred to as scouts. They are usually college students on summer vacation, but they are also high school teachers and students, and other persons, some of whom are hired on a year-round basis. These scouts are trained to identify the harmful and beneficial insects involved, to sample a field properly, and to record the field data. This training is usually provided at the beginning of each season in a two-to-four day session, with follow-up weekly training sessions during the season. In addition, continuous field training is provided by supervisory personnel working with the scouts throughout the growing season.

2.17 What are the primary duties of the field scout?

- - - - - - - - - - - - - - - -

To identify harmful and beneficial insects, sample a field properly, and record the field data.

Factors Influencing Sampling

Many factors are important to the scout and to his effectiveness in securing meaningful data. Factors such as climatic conditions, insect generation cycles, crop, stage of plant growth, soil conditions, cultural practices, surrounding habitat and adjacent fields, and time of day influence the results obtained in sampling and all must be given consideration by both the scout and the supervisor or the grower. For example, plant bug activity may be greater early or late in the day as compared with midday when the temperature is high. A change in the surrounding habitat, such as cutting an alfalfa field, may force large numbers of insects into nearby crops. Any factor that causes unusual change in insect activity or in the scout's ability to sample should be given careful consideration as to its influence on field counts.

2.18 Why should a scout note that a nearby alfalfa field has been cut?

- - - - - - - - - - - - - - - -

Large numbers of insects may have moved into the field he is sampling.

References

Anonymous. 1969. Insect-pest management and control. Nat. Acad. Sci. Publ. 1695.

Barnes, G., W. P. Boyer, J. J. Kimbrough, H. R. Sterling, and M. L. Wall. 1975. Cotton pest management program. Ark. Ext. Serv. Leaflet 52(Rev.).

Gonzales, D. 1970. Sampling as a basis for pest strategies. Proceedings of the Tall Timbers Conference on Ecological Animal Control by Habitat Management., No. 2:83-101.

Lincoln, C., G. C. Dowell, W. P. Boyer, and R. C. Hunter. 1963. The point sample method of scouting for boll weevil. Ark. Agr. Exp. Sta. Bull. 666. 31 pp.

Nielson, M. W. 1957. Sampling technique studies on the spotted alfalfa aphid. J. Econ. Entomol. 50:385-389.

Sylvester, E. S. and E. L. Cox. 1961. Sequential plans for sampling aphids on sugar beets in Kern County, California. J. Econ. Entomol. 54:1080-1084.

Wolfenbarger, D. A. and J. G. Darroch. 1965. A sequential sampling plan for determining the status of corn earworm control in sweet corn. J. Econ. Entomol. 58:651-654.

2-C ECONOMIC LEVELS

It is obvious by now that a great deal of knowledge is needed to make IPM systems work. A few of the most relevant considerations are (1) type of agroecosystem (complex diversified agriculture or simple monoculture), (2) attitude and knowledge possessed by the grower, (3) biased interests in the community and their degree of influence, (4) availability of adequately trained field personnel to sample insect populations routinely and assess their status, and possibly of greatest importance is (5) the availability and use of sound, acceptable *economic levels* (= economic thresholds) for the major insect pests on different crops in the system.

2.19 What are some of the relevant considerations for making an IPM system work?

- - - - - - - - - - - - - - - -

1. type of agroecosystem
2. attitude and knowledge of grower
3. biased interests in the community
4. adequately trained field personnel
5. sound economic levels

"Economic level" has been defined in a number of ways. The most generally accepted of these definitions is one meaning the level of a pest population where control measures must be initiated to prevent subsequent economic injury of the crop. In most cases it is differentiated from the *economic injury level*, that level which actually causes economic injury. From the grower's standpoint, however, little if any difference is assigned to the terms, as economic level means that it is time to initiate control measures if economic injury is not to result.

2.20 What is an economic level?

- - - - - - - - - - - - - - - - -

It is the level of the pest population, as determined by valid sampling procedures, at which some additional control or management practice must be used to prevent yield or quality losses.

This definition of economic level is a somewhat restricted one, and deals primarily with the pest-host plant relationship, especially with the point at which the feeding of the pest starts reducing the yield or the quality of the crop. This totally ignores other economic aspects of the problem, such as the difference in cost of control measures to the grower and the environmental costs resulting from insecticide application. The treatment here is restricted to the pest-plant association and to the point whereby yield and quality losses are inevitable. From the standpoint of implementing IPM programs, we feel that much still needs to be done in terms of the direct relationship of the plant pest to the host plant concerning the level required to cause losses in yield or quality to the grower. This is not to say that the social and environmental aspects are not important. These are real problems and should be dealt with, but at a different level from the in-field assessment of populations relative to crop losses. At the time economic levels are reached, the social and environmental aspects should be considered and should play a major role in determining the strategy selected to manage the insect pests.

It has been said that invalid estimates of economic levels are more often the rule than the exception in agriculture. The majority of such estimates are not experimentally obtained, but are based on long-standing practices, arbitrary selection, estimates used in other areas, and similar invalid criteria. In general, this assessment of the status of economic levels is too harsh. History shows that many economic levels have been raised as additional data become available for more precisely defining the levels causing damage. So the previous thresholds were invalid! They were conservative to the point of calling for insecticide treatments prematurely when compared with the new thresholds. A good case in point involves the economic level of the bollworm on cotton in the southwestern United States. With the accumulation of new research data, the treatment level of four bollworms per one hundred plants was raised to fifteen small larvae per one hundred plants. Large larvae are not counted since they cannot be controlled with available insecticides.

The economic levels—from the standpoint of the public's thinking—can be influenced by many factors. Two that probably have had opposite effects are (1) federal regulations which established limitations on subsidy payments and (2) drastic price changes on the world market. A good example illustrating both involves cotton. When cotton-subsidy payment limitations were imposed, growers were much more tolerant of pest infestations in their crop, but in 1973 when cotton prices soared, the same growers were applying insecticides at the slightest indication of pest insects in their cotton. In both cases, the concept of economic levels for various pests in cotton was ignored. Regardless of how much they treated the cotton—trying to protect every single square or boll—yields were not increased unless pest populations exceeded the economic levels.

2.21 Name two factors influencing the use of economic levels by growers.

- - - - - - - - - - - - - - - - - -

1. federal regulations, for example, marketing standards and subsidy payments
2. market prices

The risk or uncertainty of pest-associated losses, in many cases, results in preventive or insurance treatments. With additional information on the ecology of the pest to enhance predictability of damaging infestations, a more realistic approach in applying immediate control techniques, such as insecticides, can be taken.

This situation is subject to change, however, depending on such factors as the need for all the applications, the changing status of pests, market standards, long-range effects (for example, resistance and destruction of beneficial species), and availability of immediately effective alternative methods that are competitive.

To illustrate the importance of the first factor mentioned, the need for all the applications, let us consider a 1966 statistic: In the United States in 1966, farmers used 64.9 million pounds of insecticide on cotton; this amounted to 47 percent of the total insecticides used by farmers that year. Considerable research by the authors, and others as well, has shown that properly timed applications, the correct insecticide and dosage, and the use of economic levels can reduce the amount of insecticide needed by almost one half. If this were to be done on an area-wide basis, almost immediate benefits relative to IPM would ensue because of a better balance of beneficial insects to pest populations. With the accumulation of data permitting a continuous refinement of economic levels, even more precision in the use of insecticides will be possible, which in turn should permit a wider variety of options for managing pest populations.

To give another illustration, when the pink bollworm became a general problem in Arizona, growers were initiating insecticide treatments at the first sign of activity in their fields and continuing on schedule for the remainder of the season. Figures 2.4-5 indicate that some protection was needed but not to that extent. By utilizing the economic threshold of 15 percent infested bolls, insecticide applications were almost halved and yet equal amounts of cotton were produced. Now a better understanding of the population dynamics of the pink bollworm will permit a considerable reduction in the ten applications made on schedule after the economic level is reached.

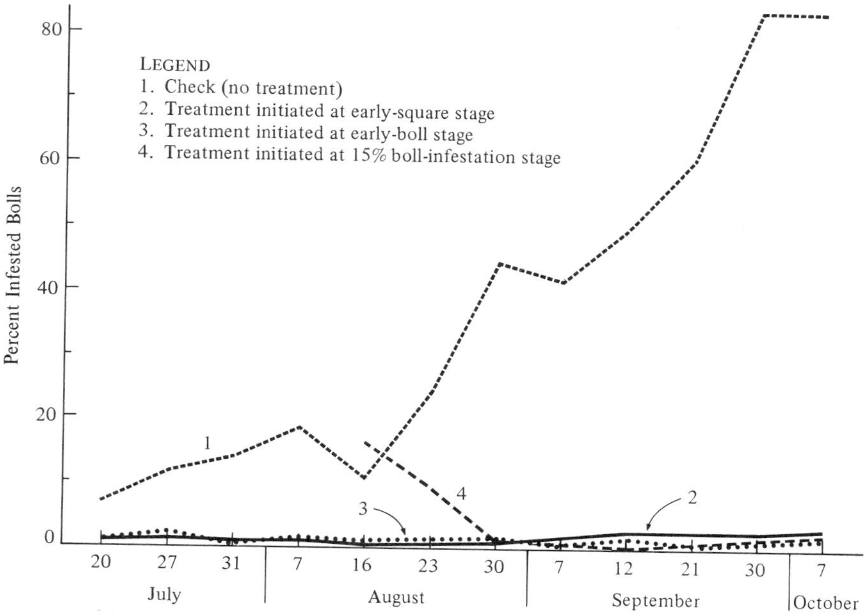

Figure 2.4 Seasonal infestation levels of the pink bollworm from three insecticide-treatment schedules and an untreated check.

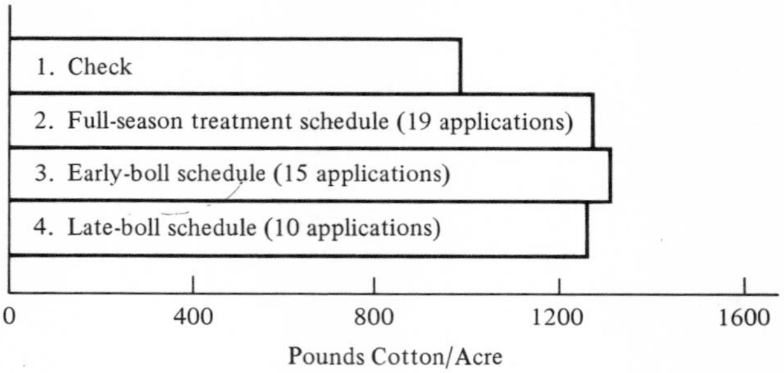

Figure 2.5 Long-staple cotton yields from three insecticide treatment schedules and an untreated check. Treatments 2, 3, and 4 were equal in yields and significantly different from the check.

The key role which economic levels play will enhance any IPM system by developing dynamic levels, incorporating such things as stage of plant growth, time of the season, crop variety, local climate, and overall crop-management practices. Even though economic levels are difficult to establish and are as complex as the agroecosystem itself, they must be constantly reevaluated in terms of continuous changes being imposed on the agroecosystem.

2.22 How can many of our present-day economic levels be improved?

- - - - - - - - - - - - - - - -

By developing dynamic levels, incorporating stage of plant growth, time of the season, crop variety, local climate, and overall crop-management practices.

References

Eichers, T. P. A., H. Blake, R. Jenkins, and A. Fox. 1970. Quantities of pesticides used by farmers in 1966. USDA, Agr. Econ. Res. Serv., Agr. Econ. Rep. 199. 65 pp.

Gonzalez, D. 1970. Sampling as a basis for pest strategies. Proceedings of the Tall Timbers Conference on Ecological Animal Control by Habitat Management. No. 2:83-101.

Smith, R. F. 1971. Economic aspects of pest control. Proceedings of the Tall Timbers Conference on Ecological Animal Control by Habitat Management. No. 3: 53-83.

Stern, V. M. 1973. Economic thresholds. Ann. Rev. Entomol. 18:259-280.

Strickland, A. H. 1970. Economic principles of pest management. Proceedings of the Conference on Concepts of Pest Management. North Carolina State University, Raleigh, N.C., Mar. 25-27, pp. 30-43.

2-D INSECT BIOLOGY AND ECOLOGY

Insects and mites are the focal point of an IPM program. Study of the insect itself will lead to answers concerning what it is, when and where it occurs, what it is doing, and its relationship to other species in the environment. The above information is essential to any type of control strategy involving economic pests in an agroecosystem, particularly when the strategy involves the complex IPM approach.

2.23 What information is needed on any insect pest?

- - - - - - - - - - - - - - - -

1. What it is.
2. When and where it occurs.
3. What it is doing.
4. Its relationship to other insects.

First, let us examine the importance of insect identification, or insect taxonomy. When a new insect pest suddenly appears on the scene, whether it is literally "new" or only new to the persons concerned, it is of utmost importance to know exactly what it is as quickly as possible. Merely placing it in a certain group of insects, such as the noctuiid family of moths or the mirid family of plant bugs, will be extremely helpful. This will provide much insight into how to start coping with the problem. Information on closely-related species can be reviewed as a starting point in unraveling its biology and ecology, and eventually its exact taxonomic relationship with the group. And where the insect poses an immediate threat to the crop, control measures successful against closely-related species may be tried. There are enough problems associated with the use of insecticides without adding to them by applying the incorrect chemical simply because you misidentified the pest. It is worse still to find an application of insecticide going on a crop to control a pest, and upon examination of the crop, to find that the "varmint" being controlled is the big-eyed bug or some other valuable predator!

Examples are available from several control methods that illustrate the importance of identifying the insect. In the biological control method, misidentification of a pest may result in the search for parasites and predators in countries other than the native home of the pest. The misidentification of the beet leafhopper as *Eutettex tenellus* Bak. led to searches for its native home in South America. It was only after P. W. Oman, a prominent taxonomist, found that this leafhopper belonged in the genus *Circulifer*, and that its native home was probably in the Mediterranean area, that biological control workers succeeded in finding several natural enemies.

In California, the biological control of several important pests was hampered for many years because of failure to recognize differences in closely related parasites in the pest's native home. Since the parasites looked alike (sibling species = morphologically identical or nearly so), better-adapted species were not introduced because they were thought to exist already in California. Once this distinction was made, proper parasites were introduced and biological control was improved.

In 1972, an outbreak of bollworms on cotton in Arizona illustrated the importance of differentiating between closely-related species. Insecticides were applied by the growers, but the expected degree of control of the bollworm was not obtained. Microscopic examination of the larvae revealed that the insect was actually the tobacco budworm, an insect known to be much more difficult to control than its close cousin. Higher dosages of the insecticides provided control. Subsequent laboratory studies substantiated the relative susceptibility of the two species (Table 2.1).

Table 2.1 LD_{50} values with lower and upper confidence limits and b values for 3 insecticide treatments applied topically to bollworm and tobacco budworm larvae. Tucson, AZ. 1972-73. (From Lentz, Watson, and Carr, 1974.)

Toxicant	Species	μg/g larval weight			
		LCL	LD_{50}	UCL	b value
Methyl parathion	*H. zea*	8.59	9.59	10.70	2.25
	H. virescens	17.72	19.70	21.89	2.56
Toxaphene : Methyl parathion (2 : 1)	*H. zea*	30.47	34.40	38.84	2.92
	H. virescens	54.08	93.88	565.48	2.15
Methomyl	*H. zea*	14.53	17.04	19.98	0.95
	H. virescens	18.54	24.77	33.09	1.05

2.24 How can correct identification of an insect aid pest management?

\- - - - - - - - - - - - - - - - -

1. It aids in biological control efforts.
2. It provides a basis for selecting insecticides if immediate control is warranted.

Another example of how a method of control might be affected by behavioral or strain differences in the insect itself involves cultural control practices directed against overwintering, diapausing larvae of the pink bollworm. In Texas the larvae overwinter in cottonseed or bolls, while in Arizona a large segment (about 50 percent) of the population leaves the bolls or seeds and spins free cocoons in the soil. Certain cultural control practices developed for Texas would certainly prove inadequate for Arizona, as it eventually turned out.

A detailed knowledge of the biology and ecology of both pests and beneficial insects is of utmost importance in any control strategy, particularly IPM. There is a direct relationship between the amount of information gathered on the total insect complex in an agroecosystem and the number of options available with which to implement IPM. Data on such important aspects as host plants, life and seasonal cycles, overwintering stage and location, plant part attacked, beneficial insects, and climatic and soil effects are only some of the essential pieces of information needed to make the wisest decisions concerning pest control or, better still, its management.

The two most important cotton pests in the United States, the boll weevil and the pink bollworm, differ considerably in their vulnerability to cultural control practices because of biological and behavioral differences. For all practical purposes both are host-specific on cotton and both attack the fruiting parts. The boll weevil spends the winter as a diapausing adult and is therefore capable of considerable movement. Most overwintering weevils seek well-drained, protective areas along fence rows and woodland for overwintering in ground trash. The problem of controlling the boll weevil culturally, therefore, would be enormous, since the weevils are dispersed over vast areas in all types of locations. On the other hand, the pink bollworm overwinters as diapausing larvae within the cotton field. Since the entire overwintering population is confined to known areas, the larval population is much more easily reduced by cultural practices than is an overwintering weevil population.

A comparison of the pink bollworm with another important lepidopterous pest, the cotton bollworm (corn earworm), illustrates the importance of other biological and ecological characteristics. Whereas the pink bollworm is relatively host-specific, the cotton bollworm is polyphagous and can be found attacking many agricultural crops. Another difference is that the corn earworm overwinters as diapausing pupae in the soil. Therefore, from the standpoint of overall survival the corn earworm should have the advantage, since it would be found in a diversity of habitats. Additionally, after emerging from overwintering quarters, the likelihood of finding a suitable host plant on which to oviposit and rear its young is much greater. Because of the oviposition behavior of the adult and the feeding behavior of the larvae, however, the pink bollworm has much greater chance for success, mainly because of escape from predators and parasites. Pink bollworm moths deposit their eggs under the tight-fitting calyx of the cotton boll, and upon hatching, the young larvae burrow into the boll where they are virtually "untouchable" by natural enemies. The bollworm moth, however, places its eggs in exposed places such as new, tender growth, and both the eggs and young larvae are vulnerable to predation and parasitism. In crops such as alfalfa, small grains, and grain sorghum, high egg and larval mortality occurs. As high as 90 percent of the corn earworm larvae collected from grain sorghum heads in southeastern Arizona have been parasitized. In cotton, predation appears to be much more important. Although egg predation on young corn silks is quite high, larval mortality from predators and parasites is generally much lower. Within the ear of corn, larvae appear to be more vulnerable to disease pathogens.

The fact that the cabbage looper is highly susceptible to a nuclear polyhedrosis virus at certain times is most useful in planning control strategies. On crops such as cotton, which can tolerate a certain amount of damage from the feeding of this pest (in other words, which have a fairly high economic level), it is almost inevitable that natural cabbage looper populations will decline ("crash" in field terminology) as a result of the disease. This usually happens during late summer in Arizona, and unless early insecticide treatments have been applied that destroyed other biological control agents, resulting in peak looper populations much earlier than usual, the disease can be used in making decisions on control. When and how this virus disease is triggered against looper populations is not known. The fact that it can and does occur naturally has been used by the grower to keep from making what would have been costly, useless, and relatively ineffective insecticide applications against this pest on cotton.

Knowledge of host plant and insect-pest relationships has been used to help control various insects that attack certain crops. Since corn is favored over cotton by the corn earworm, the staggered planting of small acreages of corn in or around cotton has served as trap crops and has prevented damage to cotton by this pest. Similarly, the strip- or block-cutting of alfalfa to maintain half-grown stands in or adjacent to cotton will reduce lygus bug damage to cotton.

2.25 How does a thorough knowledge of the biology and ecology of an insect pest influence decision-making in pest management?

- - - - - - - - - - - - - - - - -

1. It provides more flexibility in selecting an effective method of control.
2. It provides a basis for utilizing several methods in combination for management of the pest, even though any one alone might be inadequate.

Any new information on an insect pest may provide the key for even better ways of coping with it. What now appears to be fairly adequate biological and ecological data on some of the better-studied pests may in the future, with the accumulation of this new information, make today's practices seem obsolete.

We have just scratched the surface relative to manipulating the agroecosystem to make the environment less favorable for the pest or more favorable for the beneficial species, or both. The employment of IPM specialists in ever-increasing numbers should greatly accelerate the accumulation of such biological and ecological data needed to improve IPM systems.

References

Metcalf, C. L., W. P. Flint, and R. L. Metcalf. 1962. Destructive and useful insects. McGraw-Hill, New York. 1087 pp.

Schlinger, E. F. and R. L. Doutt. 1964. Systematics in relation to biological control. Pp. 247-280 in Biological control of insect pests and weeds, ed. P. DeBach. Reinhold Publishing Corporation, New York.

UNIT 3 COMPONENTS OF INSECT PEST MANAGEMENT

In the preceding units we have discussed the nature of IPM, emphasizing the principles upon which this system is based, and the basic elements essential to full utilization of this approach to insect control. We will now elaborate upon the single-component-control methods currently available and which can be incorporated into a multifaceted IPM program. These methods have been, for the most part, used individually for control of specific insect pest problems. The combining of several of these into a comprehensive IPM program can provide better suppression of key pest species and, at the same time, place less demand on any one method. The methods currently available and proved effective are cultural control, biological control, chemical control, host-plant resistance, physical and mechanical control, and regulatory control.

3-A CULTURAL CONTROL

Cultural control is one of the major control methods, predating chemical control by many years. Ironically, though older, based on sounder ecological principles, and proved effective, it was largely discarded or ignored following the appearance of the synthetic organic insecticides in the mid-1940's. Cultural control is simply the use of farming or cultural practices associated with the crop production to make the environment less favorable for survival, growth, or reproduction of pest species. In many instances it can be accomplished without additional practices, but by merely modifying those routinely performed.

3.1 What is cultural control?

\- - - - - - - - - - - - - - - - -

Cultural control is the use of cultural practices associated with the crop production to make the environment less favorable for survival, growth, or reproduction of pest species.

The most effective use of cultural control as a method in IPM requires a thorough knowledge of the life and seasonal histories and behavioral habits of the insect pest, and of its relationship to its host plants. This permits the use of cultural practices toward a vulnerable or weakened stage of the insect, or it may permit the use of such practices against the pest utilizing a behavioral characteristic or other biological peculiarity to achieve maximum effectiveness. For example, a clustering trait of the insect might allow intensive efforts to be directed against the pest when it is confined to a relatively small area.

3.2 What special knowledge is required to most effectively use the cultural method against an insect pest?

- - - - - - - - - - - - - - - -

A detailed knowledge of the pest's (1) life and seasonal histories, (2) behavior habits, and (3) relationship to its host plants.

Cultural control is seldom spectacular and may require long-term planning for greatest effectiveness, but it is often economical and dependable. This method may employ such practices as modified planting, growing, or harvesting dates; tillage operations; and crop rotation schemes. It is aimed at preventing high insect pest numbers, and associated damage, instead of destroying existing infestations. Cultural control practices are generally employed long before widespread damage is apparent. This delayed-action method of control is difficult to evaluate when compared with other methods, such as chemical control.

Many cultural control practices are based on modification of time and manner of performing certain operations and are therefore, among the most economical methods of coping with the problem. This method, however, usually does not result in complete control but merely suppresses the pest. Nevertheless, this fits much better into an IPM program because of its greater long-range stability; it is not characterized by the highly variable results obtained when chemical control alone is used.

Some of the conventional methods used in the past to achieve cultural control were sanitation, tillage, crop rotations, crop management, trap crops, and water management, some of which would logically be subtopics under the practices mentioned above. Cultural control, like other methods, has certain advantages and disadvantages. Two important advantages are that (1) many of the practices are routinely done in the production of the crop and thus do not require an additional outlay for equipment or duplication of operations to carry out insect control, and (2) they do not possess some of the detrimental side effects of insecticides, namely, insect resistance to insecticides; undesirable residues in food, feed crops, and the environment; and assault on nontarget organisms.

Probably the greatest disadvantages of cultural control are that (1) the preventative measures need to be performed long in advance of actual insect damage, (2) the preventative measures do not always provide complete economic control of insect pests, and (3) knowledge of the insect's biology and ecology by the grower is not usually adequate to permit the fullest utilization of cultural practices as a method of pest control. This last point, however, should become relatively unimportant with the employment of pest management specialists by growers or grower organizations. Under this arrangement even the first two disadvantages would be less critical.

3.3 Name some cultural practices which may be used to help control a pest.

\- - - - - - - - - - - - - - - -

1. planting and harvesting dates
2. tillage
3. rotations
4. water management
5. trap crops

3.4 What are some of the advantages and disadvantages of cultural control?

\- - - - - - - - - - - - - - - -

Advantages: (1) there is no additional outlay for equipment and (2) there are no detrimental side effects.

Disadvantages: (1) the measures need to be performed long in advance of the problem and (2) the measures do not always provide complete control.

Management of Crop Patterns to Improve Diversity

a. To enhance biological control

Cultural practices not only may act in a direct manner on the pest species, but can improve the effects of other control methods as well. In many cases, only a small change in practices is needed. A good example of this involves the grape leafhopper, a major pest of grapes in the San Joaquin and Sacramento valleys of California. This same insect, however, does not require control in the Napa and Sonoma valleys.

The eggs of the leafhopper are attacked by a native parasite, which is capable of exerting commercial control of leafhopper populations. Where the parasite is active in vineyards for an entire growing season, it effectively eliminates the third and damaging generation of the host leafhopper.

This leafhopper spends the winter in the adult stage; thus there are no eggs available for the egg parasite, and it is unable to remain in the vineyards over winter. Instead, it overwinters on a noneconomic leafhopper that breeds all year on the leaves of wild and commercial blackberries.

Vineyards in the Napa and Sonoma valleys are surrounded by blackberries, and in these valleys the grape leafhopper is not a pest requiring annual treatment. It was found that the vineyards in the San Joaquin Valley, which did have effective parasite activity, were invariably associated with some nearby source of blackberries, and the parasite population had maintained itself in the area until grape leafhopper eggs were available during the summer months.

It is apparent, then, that the grape leafhopper in California has inadvertently been made a pest by creating a barrier of time and space between the leafhopper and the overwintering source of the parasite. Whereas the leafhopper spends all winter in a vineyard, the parasite population must spend the winter on blackberries, which may be miles from the vineyards. It has since been learned that growing blackberries in or near a vineyard provides the proper grape leafhopper-parasite synchrony to effect commercial control.

b. To manipulate pest species

By strip-cutting an alfalfa field so that half of any field has young growing alfalfa available at all times, lygus bugs can be herded back and forth within the alfalfa. This prevents their movement into adjacent cotton. A more recent cultural technique has shown that interplanting strips of alfalfa within cotton fields also acts as a trap crop for lygus bugs.

Additional research has also shown the benefits of strip-cutting on certain beneficial insects. Figures 3.1-3.6 show seasonal population trends of lygus bugs and two predators in an alfalfa field where adjacent strips were alternately cut throughout the summer.

3.5 Name two advantages associated with strip-cutting an alfalfa field in a diversified agroecosystem.

- - - - - - - - - - - - - - - - - -

(1) Lygus retained in the alfalfa and (2) higher predator and parasite populations maintained in the area.

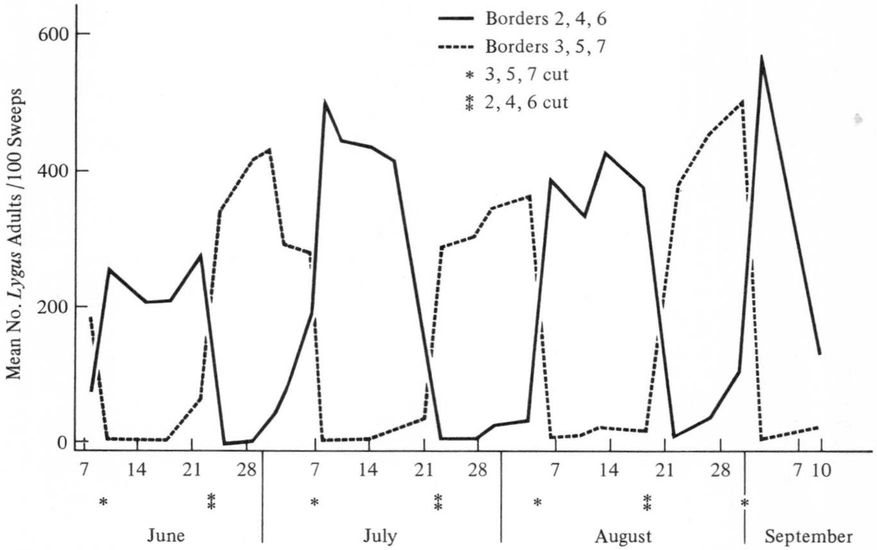

Figure 3.1 Seasonal population trends of *Lygus* adults in an alfalfa field where adjacent strips were alternately cut throughout the summer.

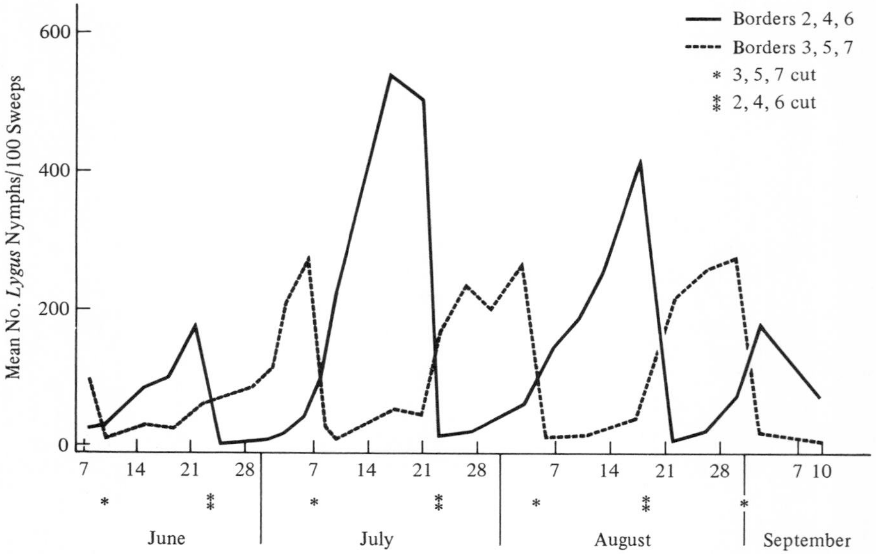

Figure 3.2 Seasonal population trends of *Lygus* nymphs in an alfalfa field where adjacent strips were alternately cut throughout the summer.

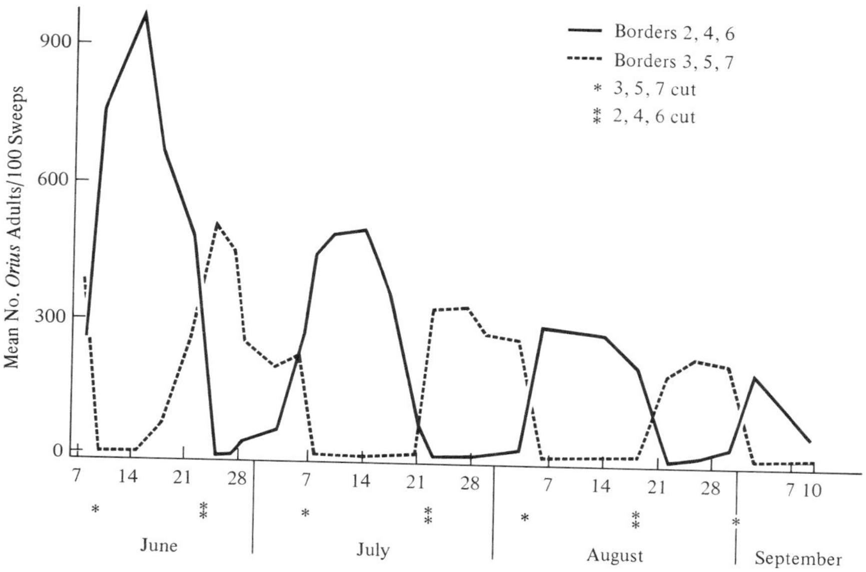

Figure 3.3 Seasonal population trends of *Orius* adults in an alfalfa field where adjacent strips were alternately cut throughout the summer.

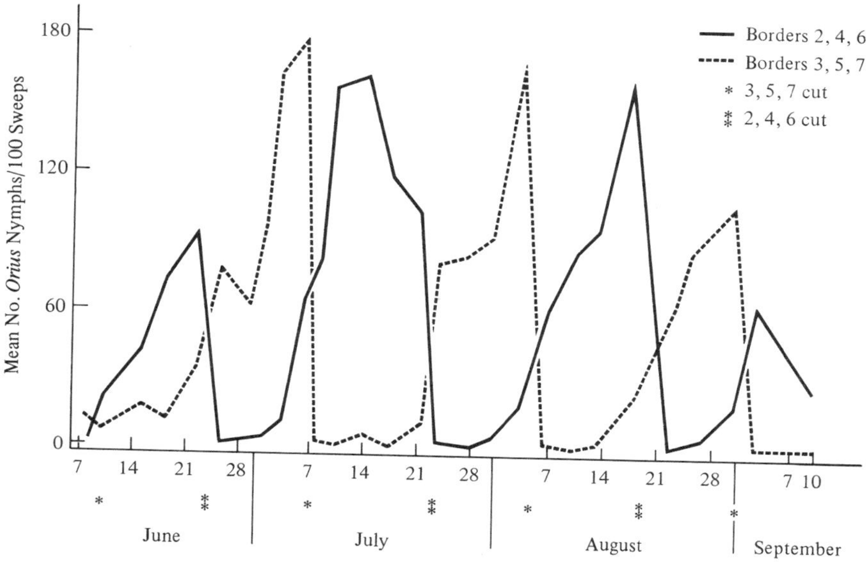

Figure 3.4 Seasonal population trends of *Orius* nymphs in an alfalfa field where adjacent strips were alternately cut throughout the summer.

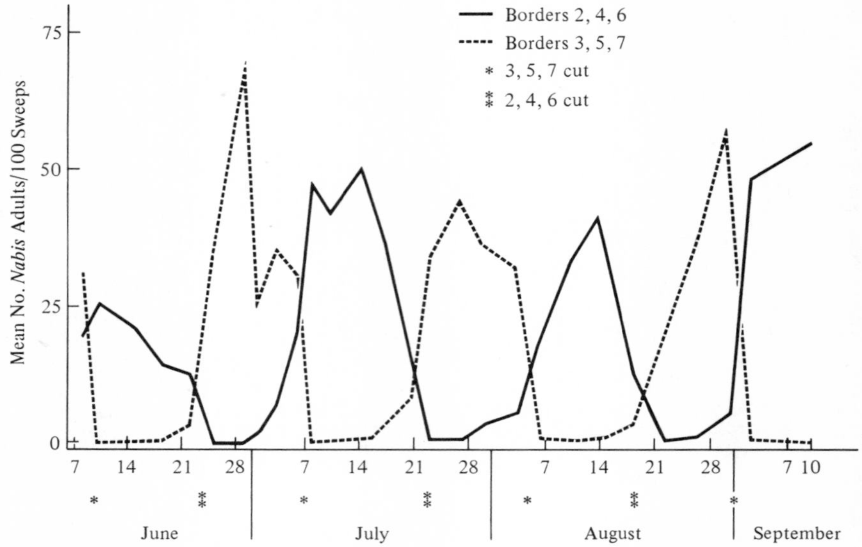

Figure 3.5 Seasonal population trends of *Nabis* adults in an alfalfa field where adjacent strips were alternately cut throughout the summer.

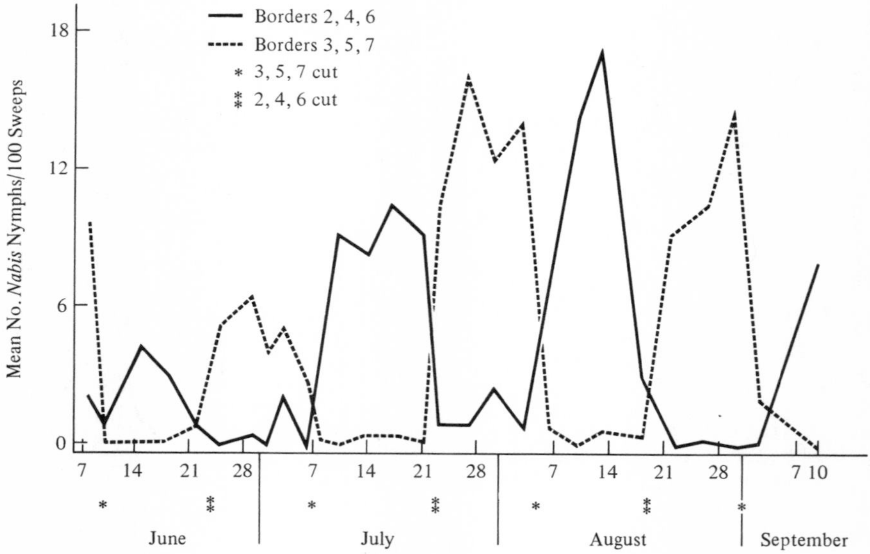

Figure 3.6 Seasonal population trends of *Nabis* nymphs in an alfalfa field where adjacent strips were alternately cut throughout the summer.

Individual Production Practices Detrimental to Pest Species

a. Tillage practices

Soil tillage practices have been effectively employed to reduce pest populations that spend a part of their life cycle in the soil. Fall plowing results in high mortality of overwintering pupae of the corn earworm, reducing the number of adults that emerge the following spring. In North Dakota the emergence of the wheat stem sawfly has been reduced by as much as 75 percent by cultivation of the stubble.

The destruction of wild or volunteer wheat is an important control measure for wheat curl mite, the vector of the virus causing wheat streak mosaic. Neither the vector nor the virus is able to live over the summer in stubble or dead plants. Consequently, summer tillage of land to be planted with winter wheat eliminates host plants and the mite vector.

b. Residue disposal

The destruction of crop residues is often a vital part of the total program to suppress serious pests. Disposal of stalk stubble by plowing under or burning has aided in the control of the European corn borer. Shredding of cotton plants in the fall of the year has aided in the control of the cotton bollworm and pink bollworm by destroying some of the insects present and by eliminating the food supply needed to build large overwintering populations.

c. Planting-harvesting dates

Insects such as the fall armyworm and the corn earworm overwinter only in the southern United States and gradually move northward during the growing season. Therefore, if corn is planted early in the northern United States, it can mature before these insects migrate northward and become economically damaging. The sorghum midge is effectively controlled in the Texas high plains and in Arizona when sorghums are planted early enough to bloom before the first week of August, since midge populations have not developed to damaging levels by this time. It is also possible to provide a host-free period by delaying the planting of a crop until egg-laying by an injurious insect has passed. Adults of the Hessian fly normally emerge in the fall and live for only three or four days. If winter wheat is planted after most of this generation has died, the plants will have few eggs laid on them. Entomologists in the Hessian fly areas have established dates for sowing winter wheat that allow the plants to make satisfactory fall growth but that are late enough to avoid heavy Hessian fly infestations.

Harvest dates can also be important. Sweet potatoes and Irish potatoes should be harvested as soon as they are mature to reduce damage by the sweet potato weevil and the potato tuberworm, respectively. Early cutting of the first and second crops of alfalfa is a practical control method for the alfalfa weevil.

d. Irrigation practices

The management of water can favor or hinder the development of insects. The greatest practical use of water management lies in the irrigated areas of the United States. Water is very important to the overwintering of the pink bollworm. Since too little or too much water is detrimental, altering irrigation practices will reduce spring moth emergence.

Example of Combined Cultural Control Practices

a. Biology of the insect and cultural control

The pink bollworm overwinters within the cotton field where it was produced, either in cotton bolls or as free larvae in ground trash and in the soil. The onset of diapause occurs in early September and increases rapidly toward the end of September and early October, reaching 85 to 90 percent by mid-October (Figure 3.7). The problem arises of how to reduce the numbers overwintering in the field prior to the first square stage of the subsequent year's cotton crop. Figure 3.8 shows a typical spring-moth emergence curve and the portion of the population emerging suicidally as compared with that emerging after squares are available for infestation.

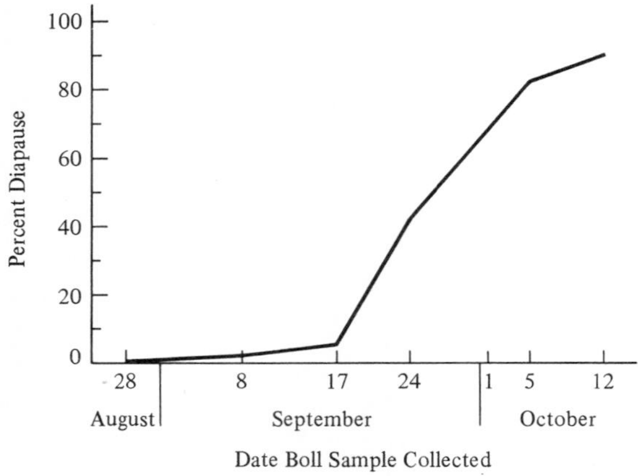

Figure 3.7 Typical diapause curve for pink bollworm larvae collected from cotton bolls.

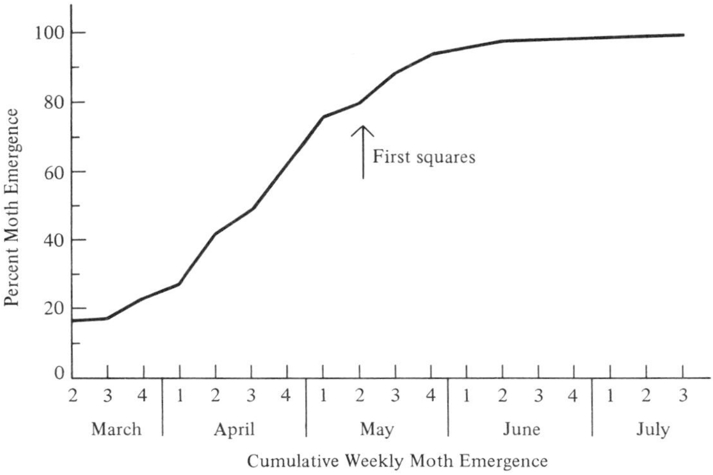

Figure 3.8 Typical spring-moth emergence of the pink bollworm. Because squares are essential for the next generation's survival, emergence prior to first squares is suicidal.

b. Cultural control practices—conventional

1. Disposal of crop residue—the practice of shredding cotton stalks has resulted in considerable direct mortality of diapausing pink bollworm larvae in Texas. Shredding of the plants in the arid southwest has not met with the same degree of success. If the stalks are removed from the field, however, about 50 percent of the overwintering population is removed also. Shredding may indirectly affect insect mortality by permitting better tillage and coverage of the larvae with soil.

2. Tillage practices—the type of tillage operation following harvest influences overwinter larval survival. In general, it appears that the more severe the tillage practice in terms of soil manipulation the greater the insect mortality (Figure 3.9).

3. Irrigation practices—the pink bollworm needs a certain amount of soil moisture for best survival. A soil moisture level of 11 to 15% seems to be best for survival. Too much moisture or too little is detrimental. It has also been shown that the level of moisture affects not only the survival but also the pattern or rate of pupation and moth emergence. Field studies have shown that the pupation and spring moth emergence patterns can be altered by the timing of irrigation water.

4. Timing of cultural practices—the time at which cultural practices are performed has an influence on their relative effectiveness. For example, deep plowing performed in January has relatively little effect on spring emergence of the pink bollworm moth when compared

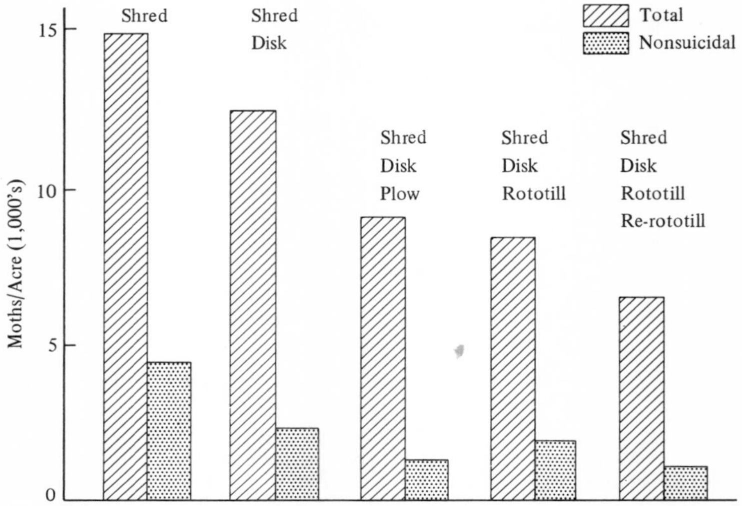

Figure 3.9 Effects of several soil tillage methods on pink bollworm moth emergence. Notice the reduced emergence from increased tillage.

with an earlier operation performed in October or November (Figure 3.10).

c. Cultural control practices—new approach

1. Early crop maturity—studies have been conducted on effects of various maturity dates of cotton on cotton yields, in-season management of pink bollworm populations, costs of producing cotton in terms of irrigations, insecticide-control costs, and so on, and the resultant levels of overwintering populations. These studies have shown that earlier crop termination than is currently practiced by growers can be achieved with less costs, no reduction in yields, and lowered spring moth emergence the following year. Figure 3.11 shows the effect of crop maturity dates on subsequent pink bollworm emergence.

2. Conventional practices, plus early crop termination—the employment of good conventional control practices such as shredding of stalks, deep plowing, and winter irrigation, in conjunction with early crop termination, should reduce the pink bollworm to a manageable level with little or no need for insecticides.

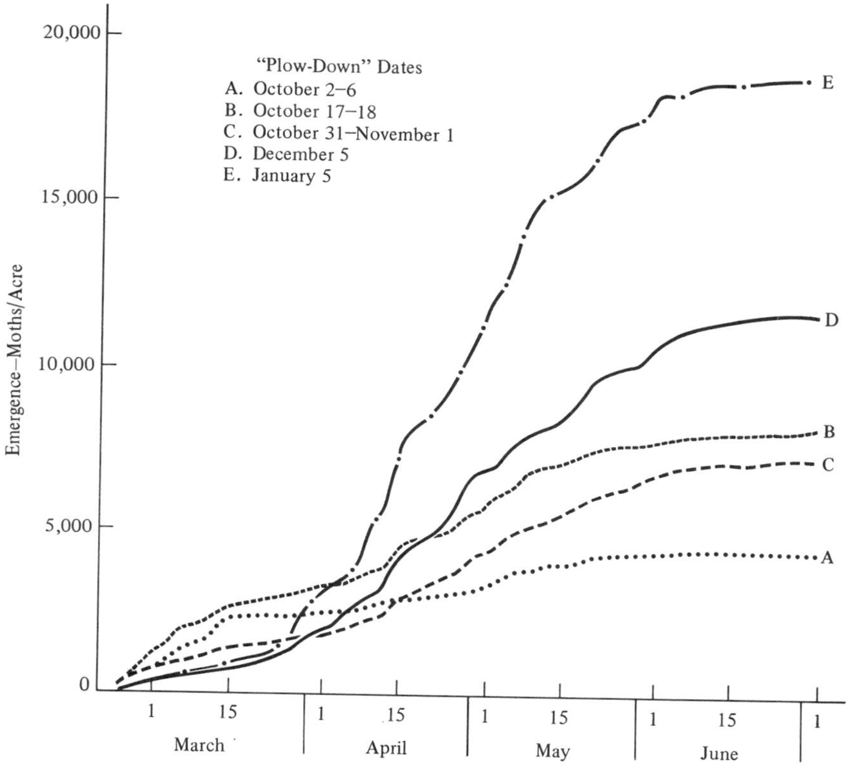

Figure 3.10 Effects of plow-down dates on pink bollworm moth emergence. Note the direct relationship of early plowing to reduced emergence.

The above discussion should leave little doubt as to the possibilities for using cultural control as one of the cheapest and least disruptive tactics in a good IPM system. Cultural control has not always been completely effective in the control of certain insect pests, but over the years it has been a significant part of crop-protection schemes. In some cases it has been the only effective means of dealing with a pest. With emphasis on IPM, it should become an even more valuable tool in managing insect pests to maintain them below economic levels. It can be incorporated into a system where other methods, such as biological control, are complementary, the sum total of which will provide satisfactory suppression of the pest species.

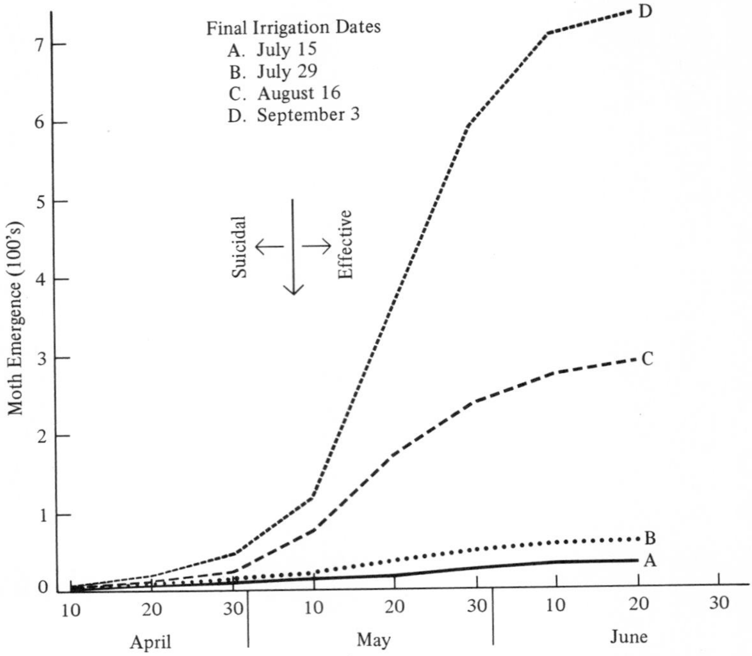

Figure 3.11 Effect of final irrigation dates and resulting crop maturity on pink bollworm moth emergence the following spring.

References

Anonymous. 1969. Cultural control. In Insect-pest management and control. Nat. Acad. Sci. Publ. 1695, pp. 208-252.

Doutt, R. L. and J. Nakata. 1965. Overwintering refuge of *Anagrus epos* (Hymenoptera: Mymaridae). J. Econ. Entomol. 58:586.

Rakickas, R. J. and T. F. Watson. 1974. Population trends of *Lygus* spp. and selected predators in strip-cut alfalfa. Environ. Entomol. 3:781-784.

Slosser, J. E. and T. F. Watson. 1972. Influence of irrigation on overwinter survival of the pink bollworm. Environ. Entomol. 1:572-576.

Smith, R. F. 1974. Management of the environment and insect pest control. Proceedings of the FAO Conference on Ecology in Relation to Plant Pest Control. Rome, 11-15 Dec. 1972, pp. 3-17.

Stern, V. M., R. van den Bosch, and T. F. Leigh. 1964. Strip cutting alfalfa for lygus bug control. Calif. Agr. 18(4):4-6.

Stern, V. M., A. Mueller, V. Sevacherian, and M. Way. 1969. Lygus bug control in cotton through alfalfa interplanting. Calif. Agr. 23(2):8-10.
Watson, T. F. and W. E. Larson. 1968. Effects of winter cultural practices on the pink bollworm in Arizona. J. Econ. Entomol. 61:1041-1044.
Watson, T. F., M. L. Lindsey, and J. E. Slosser. 1973. Effect of temperature, moisture, and photoperiod on termination of diapause in the pink bollworm. Environ. Entomol. 2:967-970.
Watson, T. F., K. K. Barnes, J. E. Slosser, and D. G. Fullerton. 1974. Influence of plowdown dates and cultural practices on spring moth emergence of the pink bollworm. J. Econ. Entomol. 67:207-210.
Wilkes, L. H., P. L. Adkisson, and B. J. Cochran. 1959. Stalk shredder tests for pink bollworm control. Tex. Agr. Exp. Sta. Prog. Rep. 2095. 2 pp.

3-B BIOLOGICAL CONTROL

Biological control, frequently condensed to biocontrol, is probably the oldest known method of insect pest control. It is recognized today as one of the most sophisticated and advanced methods of suppressing insect pests. One of the major reasons for this is that its maximum usefulness is based on sound ecological knowledge. Since biological control is applied ecology it should serve as the focal point around which other control methods are integrated. Biological control of insect pests is the reduction of pest numbers by predators, parasites, or pathogens to levels lower than would occur in their absence.

Beyond this simplified definition, there are two important views relative to biological control. One pertains to it in the ecological or "natural" sense and describes the ongoing effects of natural enemies, as when predation, parasitism or a disease epidemic has occurred. The other covers the term when human activities are concerned, thereby describing the field of biological control.

Both of these views have a place in IPM. The first is somewhat indirect and involves the careful manipulation of production and insect control practices so as not to disturb the biological control resulting from naturally occurring predators, parasites, and pathogens already in the field. The second involves human efforts to enhance biological control, either through the introduction of new natural enemies or through direct efforts to improve the efficacy of those already present.

3.6 In biological control pests are controlled by

____________________, ____________________,

or ____________________.

- - - - - - - - - - - - - - - - - -

- - - - - - - - - - - - - - - - - -

predators, parasites, pathogens

The biological control method has certain advantages over others. These are (1) safety to humans and animals, (2) permanence, and (3) economy. With regard to the economics of this method, it has been estimated that between 1923 and 1959, notably successful cases of biological control from introduced predators and parasites were of direct benefit to California's agriculture in the amount of about $115 million, while their total cost was only $4.3 million. This does not include the ensuing benefits since 1959, nor does it include the millions of dollars of benefits from programs carried out in California prior to 1923. In addition, introducing predators and parasites has reduced the use of insecticides, resulting in immeasurable benefits to the environment.

3.7 The major advantages of biological control over most other control methods are ____________, ____________, and ____________.

- - - - - - - - - - - - - - - - - -

safety, permanence, economy

Classical Biological Control

What do we mean by classical biological control? Traditionally, it means that a predator or parasite of a pest we wish to control has been collected from some distant land, returned to our quarantine laboratory for studies and rearing, and released in the desired area, with ensuing establishment and control of the pest insect. This approach is based on the fact that many of our pests were introduced, and as such, were brought in without their native parasites, predators, or pathogens. The solution, then, is to locate the native home of the introduced pest and find the natural enemies that may be holding the pest under control there. Once the parasite-host or predator-prey association is reestablished in the new area, the pest population should continue thereafter at some level lower than before. One hopes that the reduced pest population is below the economic level, and that no additional control methods will be needed. Even if the biological control agent gives only partial control, it can still be most useful in an IPM system.

3.8 Classical biological control is the reuniting of ____________ insects from the native home of the ____________ insect.

- - - - - - - - - - - - - - - - - -

- - - - - - - - - - - - - - - - -

beneficial pest

The most famous classical biological control case involved the cottony-cushion scale, and the vedalia beetle. This scale insect had virtually ruined the citrus industry in California by the late 1880's. It was eventually determined that the scale was native to Australia, and subsequent exploration in that country revealed that the vedalia beetle was in association with low populations of the scale and, further, that the beetle was a voracious feeder on the insect. In late 1888 and early 1889, three shipments of the beetle totaling 129 individuals were received in California and colonized on a caged citrus tree. Two additional shipments containing 385 beetles arrived in California in February and March 1889. From these small numbers, the beetle populations reproduced, spread, and were hand-carried to other areas, so that by late 1889 the cottony-cushion scale was completely controlled.

There are several points of importance in the cottony-cushion scale/vedalia project. First, it demonstrated how the essential features of a completely successful biological control case work: the biological control agent is easily and rapidly established, and it reduces the pest to low levels and maintains it at these low levels. Second, it established the biological control method as a major tool in the arsenal of economic entomologists. Third, it provided the justification for continued support. Fourth, it furnished the time scale for measuring the durability of biological control, meaning that it was essentially permanent in that the beetle kept the cottony-cushion scale under control until 1945 when organochlorine applications upset this balance and permitted the scale to increase once again.

3.9 What biological control project established biological control as one of the major methods of insect control?

- - - - - - - - - - - - - - - - -

The cottony-cushion scale/vedalia project.

Naturally Occurring Biological Control

Many potential pests are kept under control by naturally occurring predators, parasites, or pathogens. When these potential pests suddenly develop to damaging levels, that is, become "pests," a disturbance is indicated, human-caused or otherwise, in the normal situation. Many of us have observed insect pest outbreaks because of just such disturbances. The application of a broad-spectrum insecticide will usually destroy the beneficial insects and thus permit the insects and mites on which they feed to increase to pest status.

3.10 What is naturally occurring biological control?

- - - - - - - - - - - - - - - - -

The ongoing control of any pest by predators, parasites, or pathogens occurring naturally in the area.

Many examples could be cited showing the importance of naturally occurring biological control. An extensive study on alfalfa in California showed that a small parasitic wasp played the key role in regulating the numbers of the alfalfa caterpillar, and that without this parasite, alfalfa could not be grown economically in the central valley of California. The number of alfalfa caterpillars required to cause economic damage was determined, as were practical means of assessing populations of both the caterpillar and its parasite. It was demonstrated that when ten nearly mature, unparasitized larvae were found per sweep, biological control was not effective and other means were needed. This research, which established the economic level of the alfalfa caterpillar, determined parasite effectiveness and developed an easy sampling technique, paving the way for the development of supervised control, a forerunner of today's pest management.

For many years, the native cotton leafperforator was considered a minor pest of cotton in Arizona and southern California. There were occasional outbreaks but generally natural enemies, consisting of nine kinds of parasites and ten kinds of predators, kept it under control.

Since the spread of the pink bollworm into Yuma County, Arizona, and into southern California in 1965, scheduled applications of insecticide sprays have been made to control this pest. By the summer of 1968, the cotton leafperforator had become a serious pest of cotton in these areas and scheduled insecticide applications were required for control.

Improving Natural Enemy Effectiveness

The total environment, in addition to the unique characteristics of a particular natural enemy, determines the success of natural enemies. There are two important ways to improve their effectiveness, whether they are introduced or native. These are augmentation and conservation. The phase of biological control known as augmentation simply means the manipulation of natural enemies themselves in order to make them more effective in the suppression of pest populations. Generally, it requires detailed studies on the biology and ecology of the natural enemy to determine the type of manipulation necessary to allow it to reach its full potential as a biological control agent. There are two important methods by which the natural enemies themselves can be manipulated. These are (1) periodic colonization—the release of large numbers in an area from mass production in the laboratory or from field-collected populations, and (2) the development of better adapted strains by artificial selection.

Neither of these methods dealing with manipulation of the natural enemy offers as much promise for an ongoing pest management program as does the phase of biological control known as conservation, the modification of the environment to make it more suitable to the natural enemy. This involves everything from routine agronomic practices to the wise use of insecticides in the total pest management program.

Beneficial insects must have food, shelter, and protection from the detrimental effects of chemicals, dust, and so on, over which man has some control. These requirements can be partially fulfilled in most agroecosystems by careful attention to crop-planting patterns, rotations, harvesting practices, and the use of sound economic levels on which to base insecticide application decisions. Even where insecticide applications are needed, it may be possible to protect part of the beneficial insect complex by spot treatments, minimum dosages needed to kill the pests, or the use of an insecticide more toxic to the pests than to the natural enemies.

The effectiveness of natural enemies depends upon the degree of permanence, stability, and general favorability of the environment in which they occur. Many current agricultural practices could be modified to improve conditions for beneficial insects. These include chemical pesticide-use practices, cultivation, irrigation, crop diversification, and harvesting.

3.11 In a pest management program, what is the best method of improving the effectiveness of beneficial insects?

- - - - - - - - - - - - - - - - -

Modifying the environment to make it more favorable to the beneficial insect.

Biological Control and IPM

Relative to crop production practices, you might say that what is good for biological control is good for pest management, and vice versa. The key here is to manage the entire cropping system in such a way as to minimize the development of insect pest problems and at the same time improve the survival and efficacy of natural enemies. The cropping system should provide as many checks and balances as possible. Additionally, sound economic levels should be developed and used in all crops in the system to minimize insecticide use.

The agroecosystem is literally a gigantic insect nursery or insectary, and, unless adversely disturbed, usually produces the diversity and numbers of predators and parasites that keep many of our pests and potential pests "under control." At present, the practice of producing large numbers of predators or parasites in the laboratory for field release to control economic infestations of pest insects appears much less promising than does the practice of making the field insectary more favorable for growth, reproduction, and survival of the naturally occurring beneficial insects.

3.12 Of the two ways of improving natural enemy effectiveness—augmentation and conservation—_______________ is the most promising and practical.

\- - - - - - - - - - - - - - - -

conservation

References

DeBach, P. 1964. The scope of biological control. Pp. 3-20 in Biological control of insect pests and weeds, ed. P. DeBach. Reinhold Publishing Corporation, New york.

DeBach, P. 1974. Biological control by natural enemies. Cambridge Univ. Press, New York. 323 pp.

DeBach, P. and K. S. Hagen. 1964. Manipulation of entomophagous species. Pp. 429-458 in Biological control of insect pests and weeds, ed. P. DeBach. Reinhold Publishing Corporation, New York.

Doutt, R. L. 1964. The historical development of biological control. Pp. 21-42, ibid.

Hagen, K. S., R. van den Bosch, and D. L. Dahlsten. 1971. The importance of naturally-occurring biological control in the western United States. Pp. 253-293 in Biological control, ed. C. B. Huffaker. Plenum Press, New York-London.

Michelbacher, A. E. and R. F. Smith. 1943. Some natural factors limiting the abundance of the alfalfa butterfly. Hilgardia 15:369-397.

van den Bosch, R. and P. S. Messenger. 1973. 180 pp. Biological control. Intext Educ. Publications, New York. 180 pp.

3-C CHEMICAL CONTROL

Pesticides, chemicals used to control the harmful organisms we know as pests, play a significant role in everyday modern life. To ignore them or to hold them in contempt is to ignore the facts. Like it or not, we have isolated ourselves from the "back-to-nature" methods used in turn-of-the-century agriculture by the steady upward movements of the gross national product and our affluence; it is even virtually impossible to return to agricultural practices of the mid-1950's. There are simply not enough personnel available to plant, thin, cultivate, dust, spray, fertilize, weed, irrigate, and harvest our crops as we did then. Instead, all of these things are done in the late 1970's with powerful, human-replacing machinery, and by the generous use of those remarkable molecules we know as pesticides.

3.13 Why is it impossible to return to the agricultural practices of the 1950's?

\- - - - - - - - - - - - - - -

There are not enough personnel available to do the job.

When two out of every three persons on the earth go to bed either in a state of physical hunger or nutritional deficiency, when 70 million new mouths appear each year to feed, and when thousands are killed each year by debilitating diseases which are transmitted by insects, there are ample reasons to continue the use of insecticides. These chemicals are as much tools of agriculture and medicine, and as important, as the tractor and antibiotics. Without pesticides it is estimated that half again as much land would be required to produce the same amount of food and fiber as we presently have under cultivation. Figure 3.12 shows that cropland used for crops in the United States declined during the 1950's and 1960's but started to increase again in about 1972. During the decline, however, crop production per acre was increasing, which more than offset the reduced acreage. Part of the increase per acre yields can probably be attributed to the use of insecticides, but other factors, such as improved varieties, better agronomic practices, and planting only on the better soil, certainly were important contributions to increased production. The separation of the crop and insecticide production curves in this graph are indicative of the present overuse of insecticides.

Let us turn now to insecticides, chemicals used to control harmful and annoying insects and other arthropods. Generally, insecticides are used to protect our food and fiber plants and domestic animals, to prevent the trans-

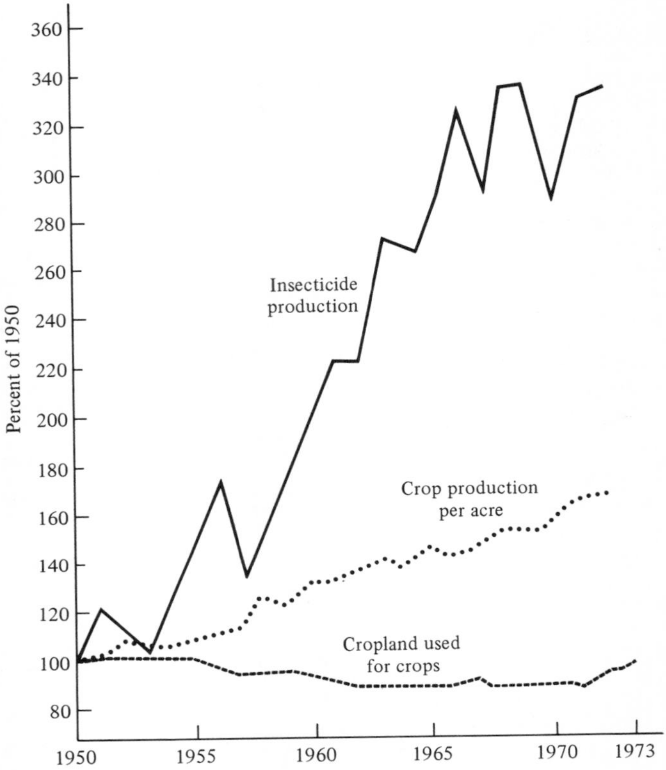

Figure 3.12 United States production of insecticides and its relationship to acres under cultivation and production per acre[1]

mission of insect-borne diseases, and to free segments of our living areas from annoying arthropods. Insecticides have become essential in maintaining our high standard of living in this "golden age" of humanity and remain the first line of defense when arthropod damage reaches economic levels.

3.14 Insecticides will remain the first line of defense in insect pest control when arthropod damage reaches ____________________________________.

\- - - - - - - - - - - - - - - -

economic injury levels

---

[1]Modified from 1974 Handbook of Agricultural Charts, United States Department of Agriculture.

Following World War II, there were some 30 insecticides in use, used at the rate of 200 million pounds per year. Today the number of registered insecticides exceeds 355, produced at the rate of 639 million pounds per year (1973). The peak years had been 1968 and 1969 when production reached 581 million pounds. The average wholesale value per pound was seventy-eight cents in 1973. Insecticides accounted for 47 percent of the tonnage of synthetic organic pesticides produced, with herbicides in second place at 38 percent, and fungicides-nematicides at 14 percent. Obviously, insecticides play a major and vital role in agricultural production.

3.15 Even though there were only about 30 insecticides in use thirty years ago, we now have more than __________ registered, making up __________ percent of the tonnage of synthetic organic pesticides produced.

- - - - - - - - - - - - - - - - -

355 49

Origins of Scheduled Chemical Applications

How did we get off on the wrong foot with schedules, calendar spraying, and early control systems that depend entirely on systematized treatments? Very likely it began as early as 1868, when kerosene emulsions were developed for the control of various scales on citrus, particularly the San José scale. Time passed and emulsions were applied to both citrus and pome fruits during the dormant growing season of the trees. Then, in 1880, lime-sulfur sprays came into use for the control of San José scale in California. Then came the discovery, in 1902, that lime-sulfur was effective against apple scab. The use of these materials as scalicides and fungicides on a regular annual basis established a comforting facet in the use of chemicals for pest control: regularity. Disease and scale populations would no longer flare up as they had in the past. Instead, they could be held at low, noneconomic levels with a single, mid-winter application of spray. Effective controls were just becoming available at minimal cost.

3.16 Our dependence on systematic applications of pesticides probably began on the crops of ______________ and ______________.

- - - - - - - - - - - - - - - - -

citrus and apples

In 1918, calcium arsenate dusts were found effective in controlling boll weevils on cotton, and that chemical soon became the insecticide of choice. Here it was demonstrated that the proper use of these dusts gave excellent control of the weevil with acceptable increases in cotton yields, setting the stage for movement of applied entomology into the excessive reliance on the use of insecticides.

Shortly after the introduction of calcium arsenate for control of the boll weevil on cotton, the airplane was brought into the insect control picture. In 1921, a bi-wing plane was used to make the first aerial application of calcium arsenate dust for control of the catalpa sphinx in Troy, Ohio. With this new method of application it was soon demonstrated that insect-control chemicals could be applied to 50 acres in less than an hour by one man. Aerial application of insecticides did not become common place, however, until after World War II with the availability of large numbers of pilots and war-surplus planes.

3.17 What two developments brought systematic applications of insecticides to field crops?

\- - - - - - - - - - - - - - - - - -

Calcium arsenate dusts and aerial application.

Calcium arsenate, however, was not the answer to all insect control problems on cotton. When it was used on entire fields, populations of predators such as lady beetles, ground beetles, and lacewings were severely affected. The loss of predators was usually followed by serious outbreaks of aphids and, frequenly, bollworms. The losses resulting from the activities of these consequential pests were often greater than were the benefits obtained from controlling the boll weevil.

The calcium-arsenate era came to an end when the organochlorine insecticides appeared after World War II, beginning in 1946. These were DDT, BHC, toxaphene, chlordane, aldrin, heptachlor, dieldrin, and endrin. For the first time in human history, cheap, effective insecticides were available. Their use immediately produced visible results. Growers were quickly convinced that all their pest problems were at an end because of these very effective and economical insecticides. It could be said that the organochlorines were used with gusto not only in cotton insect control but on practically all crops from 1948 until the 1960's. During this interval the sterile field, one devoid of insects, was the epitome of plant protection, with the early determinant concept for chemical control, economic levels of insect pests, being disregarded completely.

Even though many agricultural crops rely heavily on insecticides, only a few are responsible for a majority of the tonnage released into our environment each year. Table 3.1 shows that cotton utilizes 47 percent of the insecticides applied to all agricultural crops. This is followed by corn which uses 17 percent of the crop insecticides. A concentrated effort to utilize the pest-management approach would drastically reduce the amount of insecticides used on these two crops, and thus would significantly reduce the total tonnage used on all agricultural crops. Undoubtedly, the pest-management approach in the other crops would result in a reduction of the amount of insecticide used on those crops as well.

Table 3.1 Insecticide usage on agricultural crops of the United States.*

Crop	Acres treated (percent)	Amount of agricultural insecticides used (percent)
Nonfood	1	50
Cotton	54	47
Tobacco	81	3
Food	4	NA
Field crops	NA	NA
Corn	33	17
Peanuts	70	NA
Rice	10	NA
Wheat	2	NA
Soybeans	4	2
Pasture, hay and range	0.5	3
Vegetables	NA	8
Potatoes	89	NA
Lettuce	100	NA
Cole crops	100	NA
Fruit	NA	13
Apples	92	6
Citrus	97	2

*Adapted from Integrated Pest Management, prepared by the Council on Environmental Quality, November 1972.

The Benefits

Benefits resulting from this lavish use of organochlorine insecticides for the most part could be described in four words: economy, speed, persistence, and yields. Thanks to the ready availability of these new synthetic insecticides, which cost from twenty-five to seventy-five cents per pound before formulation into dusts and sprays, insect control was simple, easy, and economical. Anyone could apply these materials in practically any manner and achieve spectacular results. A new philosophy was born and has prevailed from that time: "The only good bug is a dead bug!"

3.18 What were the four prime benefits of depending entirely on insecticidal control?

- - - - - - - - - - - - - - - -

Economy, speed, persistence, and yields.

In cotton insect control, a technique known as early control of boll weevils, thrips and other insects was developed. Two applications of the chlorinated insecticides were applied eight to ten days apart to cotton at the first appearance of the very small or pinhead squares. This program caught the attention of cotton growers, and soon developed into one of applying insecticides to cotton on a regular schedule throughout the season without regard to pest populations or economic levels. Corn, which had been plagued over the years in the north central United States with the northern corn rootworm, was suddenly freed of this pest with the application of dieldrin or heptachlor to the soil to control the damaging larval stages. Soon all fields were treated every year with the organochlorines at planting time—just as a matter of insurance.

Apple and other fruit industries had already developed schedules for applying oil emulsions in the dormant season to control scales and mites before the new chlorinated insecticides appeared, so it was a simple matter to carry insect control throughout the growing season, adhering to a scheduled application system. Once the schedules were developed, seven or eight sprays were applied automatically, beginning with the first pre-bloom spray. Again, the pesticide-use pattern had become based on the calendar and the biological clock of the fruit tree, without regard to pest populations or economic thresholds.

Vegetable production joined the same trend. Worms, worms, and more worms were the perpetual pest of vegetables, and worms could be controlled with the organochlorine insecticides. Within no time, vegetable crops were automatically being treated on schedule with fungicides and insecticides to control diseases and insects. Besides, the finicky housewife much preferred fruit and vegetables with no disease blemishes or worm holes.

Citrus growers readily accepted this easy and economical method of insect control—scheduled applications.

Thrips, scale insects, and mites all succumbed to the new materials. The blemish-free orange, grapefruit, and lemon were now a reality. Again, housewives preferred perfect fruit—no scars, no blemishes.

And so it went. Insect control was now the simplest of all the growing operations, requiring only application equipment and the pesticides, and no particular knowledge or skills. It became a part of the built-in planning of growers, as did fertilization, cultivation, and harvest. With almost total insect control amounting only to \$10 to \$20 per acre, who could afford to run the risk of "letting them eat me up?"

But why should insecticides not be used abundantly? See how they have increased yields: Cocoa production in Ghana, the largest exporter in the world, has been trebled by the use of insecticides to control just one insect species; and Pakistan sugar production was increased 33 percent through the use of insecticides. The Food and Agricultural Organization (FAO) has estimated that 50 percent of the cotton production in developing countries would be destroyed without the use of insecticides. In the United States there were some equally good responses in terms of increased yields: cotton, 100 percent; corn, 25 percent; potatoes, 35 percent; onions, 150 percent; tobacco, 125 percent; beet seed, 180 percent; alfalfa seed, 160 percent; and milk production, 15 percent. Insecticides have thus become indispensable to today's advanced agricultural technology.

3.19 Why have insecticides become indispensable in today's agriculture?

\- - - - - - - - - - - - - - - -

To maintain economic yields.

3.20 In your own words, why has the pattern of total dependence on insecticides developed for most of the major crops?

\- - - - - - - - - - - - - - - -

Becoming a part of our overall mechanized agriculture, the automatic, preventive form of pest control was readily accepted because it was easy to schedule and economical.

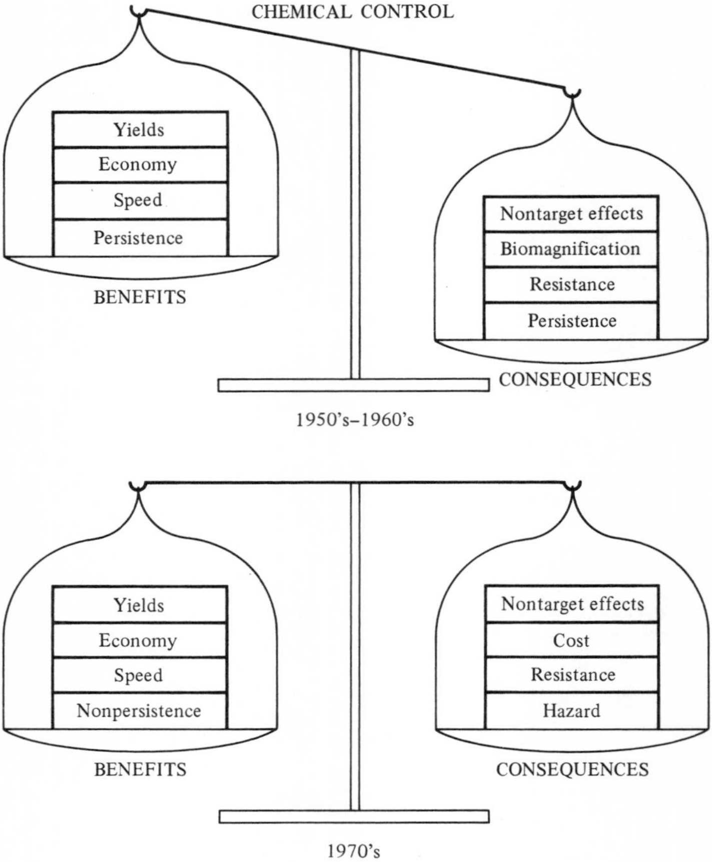

Figure 3.13 Diagrammatic illustration showing the importance of considering both the benefits and consequences when using insecticides. Compare the 1950's-60's with the 1970's as influenced by insecticides available during these periods.

The Consequences

The consequences of most concern, as we reflect on the use of organochlorine insecticides in the twenty-five years from 1946 through 1970, are persistence, biomagnification, resistance, and toxicity to nontarget organisms. Some of these consequences, particularly resistance and effect on nontarget organisms, also result from the use of the organophosphates and carbamates. In making a decision to use a chemical insecticide to control a particular pest, many factors must be considered, particularly the benefits and consequences of applying the chemical. The benefits must at least equal (balance out) the consequences and should preferably exceed them (Figure 3.13).

a. Persistence and biomagnification

Persistence is the quality of a compound to retain its chemical identity and biological effectiveness for long periods of time. This is considered highly desirable for continued control, but it also causes some environmental problems. A case in point is DDT.

DDT was probably the most useful and economical insecticide available to humanity. Consequently, it was used not only in great quantity, but in excessive quantity. Its persistence and fat-solubility permitted it to accumulate in animals or plants and enter the food chain of both humans and wildlife. Consequently, humans and wildlife at the tops of these food chains received large exposures to DDT, and its equally stable metabolite DDE, simply through ingestion of food, an accumulation termed *biomagnification*. A good example of increasing concentrations of some chlorinated insecticides owing to this magnification is shown in Table 3.2.

Table 3.2 Concentrations of organochlorine insecticides in Lake Michigan*

Substrate	Residues (ppm)
Bottom sediment	0.0085
Small invertebrates	0.41
Fishes	3.0-8.0
Herring gulls	3177

* Hickey et al. 1966.

3.21 Biomagnification occurs when one species of animal eats smaller ________________ that contain ________________ insecticides or their metabolites.

- - - - - - - - - - - - - - - -

animals persistent

3.22 The two characteristics of DDT that permit it to accumulate or undergo biomagnification are its ________________ and ________________.

- - - - - - - - - - - - - - - -

- - - - - - - - - - - - - - - -

persistence fat solubility

Persistence and biomagnification are particularly important when the compound in question is capable of causing biological effects, for example, when it is acutely toxic, carcinogenic, teratogenic, or mutagenic, or when it is capable of causing other chronically toxic conditions.

Persistence and biomagnification have been the basis for cancelation of a number of registrations of the organochlorine insecticides by the EPA. It is apparent that potential environmental effects must be considered carefully in the design and later use of new insecticides.

3.23 Most agricultural registrations of the organochlorine insecticides have been canceled by the EPA because of two important characteristics, ____________ and ____________.

- - - - - - - - - - - - - - - -

persistence biomagnification

The two areas of concern in this manual regarding chemical control, which are also the two consequences that have had the most influence on promoting IPM directly and indirectly, are *toxicity to nontarget organisms* (namely, beneficial parasites and predators) and *insect resistance to insecticides*.

b. Resistance

Resistance or tolerance is now a common term: most staphylococcus or staph infections are resistant to penicillin; flies have developed resistance to DDT; cockroaches have become resistant to chlordane; some strains of gonorrhea cannot be controlled with any but the very latest antibiotics; and some recently developed fungicides with specific modes of toxic action fail to control certain plant diseases.

The development of resistance to insecticides by insect pests dramatically demonstrates a mini-form of evolution. The susceptible insects are killed, leaving behind only those that are genetically resistant to the insecticide. Resistant individuals make up an increasingly large part of the pest population and pass their resistance on to future generations. Resistance in insect pests simply represents the survivors of a stringent biological selection mechanism, the insecticide, over several generations. The greater the number of generations exposed to an insecticide in a period of time, the greater the potential for developing resistance as a consequence of this intensive chemical selection.

3.24 What increases the potential for development of resistance by an insect?

- - - - - - - - - - - - - - - -

Increased exposure to an insecticide.

Insect resistance to insecticides is a problem worthy of genuine concern. In 1944, only 44 insect species were known to have developed resistance to various insecticides in some populations. Remember that the new synthetics were not yet on the scene. Today's estimates place this number higher than 250, half of which are of agricultural significance.

As more insecticides become ineffective against certain species, the problems of insect pest control increase proportionally. Populations of several insect pests now have such a high proportion of individuals resistant to all known insecticides that substitute materials are no longer available and insecticides are not recommended as control measures. We are literally exhausting our arsenal of chemical tools. Public health problems from disease vector resistance are likely of greater importance, but will not be considered in this book.

3.25 Why is insect resistance to insecticides a concern of pest management?

1. ______________________________
2. ______________________________

- - - - - - - - - - - - - - - - - -

1. The number of resistant species is growing rapidly.
2. We are running out of new insecticides.

One of the less subtle phases of insecticide resistance is cross resistance. Insects that have developed resistance to one organochlorine insecticide are frequently resistant to another organochlorine, even though they have not been previously exposed. In many instances, insects that became resistant to the organochlorines quickly became resistant to the organophosphate insecticides. And because the modes of action of the organophosphates and carbamates are similar, insects that became resistant to the organophosphates soon became resistant to one or more of the carbamates.

3.26 What is the hidden danger from cross-resistance?

- - - - - - - - - - - - - - - - - -

Insects may be developing resistance to compounds to which they have never been exposed.

These examples of resistance can be attributed to the singular method of action of a particular insecticide, which disrupts only one genetically controlled process in the metabolism of the insect. The result is that resistant populations appear suddenly, either by selection of resistant individuals in a population or by a mutation, which appears to be the less common of the two routes of resistance development. Generally, the more specific the site and mode of insecticidal action, the greater the likelihood for an insect to develop a resistance to that chemical.

The story we are about to retell has been repeated the world over, in one pest situation after another. It is a typical example of the development of resistance in an insect raised by insecticidal pressure from a seat of secondary importance to one of primary importance. In the Cañete Valley of Peru, prior to 1949, cotton insects were controlled with arsenicals and nicotine sulfate; cotton yields were 470 pounds per acre annually. However, in 1949, heavy outbreaks of the cotton bollworm and aphids occurred, and cotton yields decreased to 326 pounds. Between 1949 and 1956 Cañete growers relied totally on DDT, BHC, and toxaphene; cotton yields doubled. At the same time the natural insect enemies were depleted. Then these chlorinated insecticides all lost their effectivness as the bollworms became resistant. Despite increased frequencies and rates of application, the pest insects won, and the 1955-56 growing season was a disaster. In subsequent years, an integrated control program consisting of field sampling of pests, better utilization of beneficial insects, and application of insecticides only when needed was successfully introduced, and yields quickly climbed to more than 700 pounds per acre.

Resistance has not proceeded at the same pace in all major pest species, but resistance does appear to be dangerously near in some of the most important insects in this country: the boll weevil, the cotton bollworm, the tobacco budworm, the sugarcane borer, and the rice water weevil. One very serious aspect of resistance is that when it does hit, there may be no satisfactory substitute insecticide.

The seeming ineffectiveness of an insecticide does not indicate with certainty that the insect is resistant. Effectiveness can also be reduced by the destruction of natural controls, for example, parasites and predators. Only a very small fraction of the total number of insects are considered pests. When at normal population densities, most insects pose no threat to cultivated crops, and many are important to the health and stability of the environment because they control other potentially damaging species.

3.27 Control failures may be interpreted as resistance when the real problem may be destruction of ______________________.

\- - - - - - - - - - - - - - - -

natural enemies

Most of the insecticides in use today have broad-spectrum effects. That is, they are lethal to a wide range of insects and other invertebrates, including beneficial competitors, predators, and parasites of the target pest insects. When insect pest populations are drastically reduced, their natural enemies are reduced even more. A resurgence in the pest population can then occur, resulting in increased damage to the crop.

With the use of these broad-spectrum insecticides, insects that were controlled naturally are sometimes caused to increase to such numbers that they become pests. This is the result of killing the insect's natural enemies. One quickly sees that an insect can be made a pest by "proper" use of insecticides applied to control a primary pest. In this context, this "proper" use could become an "improper" use.

3.28 The four consequences of overuse of insecticides are ______________, ______________, ______________, and ______________.

\- - - - - - - - - - - - - - - -

persistence, biomagnification, resistance, toxicity to nontarget organisms

---

c. Effects on nontarget organisms

Insecticides are generally applied to an agricultural crop for only a few pest insects. In most cases, these few "key" species require this kind of artificial control measure to prevent economic losses to the crop. The history of insecticide usage, however, illustrates the fact that additional problems are created, either by the rapid resurgence of the treated pest population or by raising minor pests to the role of secondary or major pest status.

3.29 Insecticide usage may create additional problems such as resurgence of the ______________, or raising minor pests to the status of ______________.

\- - - - - - - - - - - - - - - -

pest population	major pests

The outbreak of secondary pests or the resurgence of treated species following repeated applications of insecticides has generally been attributed to the destruction of their natural enemies. This may, or may not, be the primary reason for such changes in the pest complex. In all fairness to the insecticide method of insect control, it should be noted that some pests, formerly considered important, have declined or become almost nonexistent in the face of repeated applications of insecticides, particularly the organophosphates. For example, two formerly important cotton pests, the cotton leafworm and the cotton aphid have declined in pest status following the use of organophosphates for control of organochlorine-resistant boll weevils. This happened even though the mixtures of organochlorines and organophosphates used for cotton insect control in the southern United States since 1965 should have had equal or even more drastic effects on predators and parasites than did the organochlorines used prior to that time. The reasoning for the decline in pest status of these two insects was that the organophosphate insecticides were so effective for leafworm and aphid control that they were unable to develop populations that reached economic levels.

Other cases could be cited where certain pests declined in importance following the use of insecticides. It has been apparent for a number of years, however, that the reverse is more generally the case, that is, formerly minor pests have become major pests, and there is much evidence indicating a direct correlation with the use of insecticides. Without question, the suppression of natural enemies by insecticides is the main reason for these topsy-turvy shifts.

It may be argued that insecticides affect both pests and natural enemies alike and that the capacity of the pest species to "resurge" should apply to the natural enemy populations as well. This is probably true in many cases, and, in fact, some beneficial insects and mites (lady beetles and phytoseiied mites) have developed resistance to insecticides, along with the pest species. However, natural enemy populations are generally dealt a double blow by insecticide applications. First, they are subject to intoxication, along with the pests, by the direct application of insecticide, and thus are reduced, along with the pests, to relatively low numbers. Second, the surviving natural enemies are left with little to feed upon after the applications and may further decline because of starvation or inability to find suitable hosts for reproduction before pest resurgences occur. This can be as detrimental to natural enemy populations as the insecticide.

3.30 Natural enemies receive the old "one-two" from an insecticide application:

1. ______________________________

2. ______________________________

- - - - - - - - - - - - - - - - -

1. They are killed outright.
2. They are deprived of suitable hosts for reproducing.

Insecticides reduce natural enemies with a resultant increase in pest populations. There are two good examples to illustrate this point. In 1946 and 1947, many thousands of acres of California citrus developed damaging populations of the cottony-cushion scale following DDT applications for other pests. This was due to the elimination of the vedalia beetle, a predator which had been the main factor controlling this scale since introduction of the beetle into California in 1888. Since then it has been demonstrated experimentally that elimination of the vedalia beetle was the sole cause of the increase of cottony-cushion scale.

The second example which proves conclusively that elimination of a predator may permit resurgence of a pest species involves the cyclamen mite on strawberry plants and the predatory mite *Typhlodromus reticulatus* (Oudemans). Scientists have shown that elimination of the predator with parathion treatments permits cyclamen-mite population increases of fifteen to thirty-five times, whereas cyclamen-mite populations continue to decline where the predaceous mite is left undisturbed. Similar results can be obtained in the greenhouse when predators are removed by hand rather than with the insecticide. In both instances the removal of an effective predator permits rapid increase of the pest.

3.31 The key pest species may resurge following the use of chemical control resulting from the removal of ________________ and ________________.

- - - - - - - - - - - - - - - - -

predators parasites

In summary, even though both pests and natural enemies respond similarly to insecticides, the evidence is conclusive that some of our pest problems are man-made by the wide-scale use of insecticides. Natural enemy suppression is a major cause of changes in pest status and resurgence of treated species. It is important to exercise a conscious effort to use insecticides wisely to minimize the adverse effects on beneficial insects.

3.32 In your own words, what is meant by a "man-made pest"?

References

Brown, A. W. A. 1968. Insecticide resistance comes of age. Bull. Entomol. Soc. Amer. 14:3-9.

DeBach, P. 1947. Cottony-cushion scale, vedalia and DDT in central California. Citrograph 32:406-407.

DeBach, P. and B. R. Bartlett. 1951. Effects of insecticides on biological control of insect pests of citrus. J. Econ. Entomol. 44:372-383.

Hickey, J. J., J. A. Keith, and F. B. Coon. 1966. An exploration of pesticides in a Lake Michigan estuary. J. Appl. Ecol. 3(suppl.):141-154.

Hoffman, C. H. 1971. Restricting the use of pesticides—what are the alternatives? Pp. 21-30 in Economic research on pesticides for policy decision making. U.S. Department of Agriculture, Washington, D.C.

Huffaker, C. B. and C. E. Kennett. 1953. Developments toward biological control of cyclamen mite on strawberries in California. J. Econ. Entomol. 46:802-812.

Newsom, L. D. 1967. Consequences of insecticide use on non-target organisms. Ann. Rev. Ent. 12:257-286.

Newsom, L. D. 1974. Pest Management: History, current status, and future progress. Pp. 1-8 in Proceedings of the Summer Institute on Biological Control of Plant Insects and Diseases, ed. F. G. Maxwell and F. A. Harris. University Press of Mississippi State University, Mississippi State, Miss.

Smith, R. F. 1969. Integrated control of insects, a challenge for scientists. Agr. Sci. Rev. USDA 1st Qtr:1-5.

U.S. Department of Agriculture. 1975. The Pesticide Review 1974. Agricultural Stabilization and Conservation Service, Washington, D.C. 20250. 58 pp.

Insecticide Formulations

After the technical grade insecticide has been manufactured, it is formulated into a usable form for direct application, or for dilution followed by application. The formulation is the form in which the pesticide is sold for use, under the formulator's brand name, or it may be custom-formulated for another firm.

3.33 The technical grade insecticide is produced by the ________________.

- - - - - - - - - - - - - - - -

basic manufacturer

3.34 An insecticide is ________________ into its usable form after manufacture.

- - - - - - - - - - - - - - - -

formulated

This formulating is the processing of an insecticidal compound by any method which will improve its properties of *safety*, *storage*, *handling*, *application*, and *effectiveness*. The term is usually reserved for commercial preparation prior to actual field use and does not include the final dilution in application equipment.

3.35 Formulation of an insecticide improves its ____________, ____________, ____________, ____________, and ____________.

- - - - - - - - - - - - - - - -

safety storage handling application effectiveness

A pesticide must be effective, safe, easy to apply, and generally economical to be acceptable for use by the grower or commercial applicator. Insecticides must be formulated into many usable forms for satisfactory storage, for effective application, for safety to the applicator and the environment, for ease of application with readily available equipment, and for economy. Owing to the chemical and physical characteristics of the technical grade pesticide, this is not always simply accomplished. For example, some technical grade materials are liquids, others are solids; some are stable to air and sunlight, others are not; some are volatile, others are not; some are water soluble, some are oil soluble, while others may be both water and oil insoluble. These characteristics pose problems to the formulator, since the final formulation must meet the specifications of federal and state regulations and satisfy standards of acceptability by the user.

3.36 Explain the need for various formulations of pesticides.

- - - - - - - - - - - - - - -

The stability and physical condition of the technical material and its physical and chemical properties require various formulations.

More than 99 percent of all insecticides used in agriculture in the mid-1970's are sold in the formulations shown in Table 3.3. Familiarity with the more important formulations is essential to the well-read "pest-management specialist." We will now examine the major insecticide formulations used in agriculture.

Table 3.3 Common Formulations of Insecticides.*

---

1. Sprays
   a. Emulsible concentrates (EC)
   b. Water-miscible liquids (S)
   c. Wettable powders (WP)
   d. Flowable suspensions (F)
   e. Water-soluble powders (SP)
   f. Ultra-low-volume concentrates (ULV)
2. Dusts (D)
   a. Undiluted toxic agent
   b. Toxic agent with active diluent, e.g. sulfur
   c. Toxic agents with inert diluent, e.g. pyrophyllite
3. Granulars (G)
4. Soil fumigants
5. Baits (B)
6. Animal systemics
7. Fertilizer-insecticide combinations (FM)
8. Encapsulated insecticides

---

*This list is incomplete, containing only the more common formulations used in agriculture.

Sprays

<u>Emulsible concentrates (EC)</u>. Trends in formulation shift with time and need. Traditionally, insecticides have been applied as water sprays, water suspensions, oil sprays, dusts, and granules. Sprays are the most popular methods of application. Consequently, 75 percent of all insecticides are applied as sprays. The bulk of these are currently applied as water emulsions made from emulsible concentrates.

3.37 The majority of insecticide applications are applied as _______________ and the bulk of these are made from __________________________ formulations.

\- - - - - - - - - - - - - - - -

sprays emulsible concentrate

Emulsible concentrates, or emulsifiable concentrates, are concentrated oil solutions of the technical grade material containing an emulsifier to permit the concentrate to mix readily with water for spraying. The emulsifier is a detergent-like material that causes the suspension of microscopically small oil droplets in water, forming the emulsion.

3.38 The oil solution of an insecticide and emulsifier are the _____________________________ formulation.

\- - - - - - - - - - - - - - - -

emulsible concentrate

3.39 The ______________________ makes possible the mixing of an emulsible concentrate and water.

\- - - - - - - - - - - - - - - -

emulsifier

When emulsible concentrates are added to water, the emulsifier causes the oil, if agitated, to disperse immediately and uniformly throughout, producing an opaque liquid. This oil-in-water suspension is a normal emulsion. Recently systems of converting insecticide sprays to invert emulsions, water-in-oil, have been developed which are considerably thicker and less likely to drift than normal emulsions. Additional developments include extended residual activity, resistance to weathering, and higher rates of on-target deposit.

3.40 Suspensions made of oil-in-water are ________ emulsions.

- - - - - - - - - - - - - - - -

normal

3.41 Suspensions in which water droplets are distributed throughout oil are ___________ emulsions.

- - - - - - - - - - - - - - - -

invert

If properly formulated, emulsible concentrates should remain suspended without further agitation for at least twenty-four hours after dilution with water. If precipitates form, resulting in clogged nozzles and uneven application, additional emulsifier can be obtained from the formulator and added to the concentrate at the rate of 1 to 1.5 pounds per gallon of out-dated concentrate.

3.42 Emulsions should remain suspended with no precipitate for at least ______________________.

- - - - - - - - - - - - - - - -

twenty-four hours

Water-miscible liquids (S, for solution). Water-miscible liquids are totally miscible in water. The technical grade material may be water miscible initially, or it may be dissolved in alcohol, making it water miscible. These formulations resemble emulsible concentrates in viscosity and color, but remain clear when diluted with water.

3.43 Water-miscible liquids differ in appearance from emulsions after dilution by remaining ________.

- - - - - - - - - - - - - - - -

clear

Wettable powders (WP). Wettable powders are insecticide impregnated dusts containing a wetting agent to facilitate the mixing of the powder with water before spraying. The technical material is added to the inert diluent (in this case a finely ground talc or clay) in addition to a wetting agent similar to a dry soap or detergent, and mixed thoroughly in a ball mill. Without the wetting agent, the powder would float when added to water, and the two would be almost impossible to mix. Because wettable powders usually contain from 50 to 75 percent of clay or talc, they sink to the bottom of the spray tank unless the liquid is agitated constantly.

3.44 Three ingredients, ______________, ______________, and ______________, are used when formulating wettable powders.

- - - - - - - - - - - - - - - -

dust pesticide wetting agent

Flowable suspensions (F). Flowable suspensions are an ingenious solution to a formulation problem. Earlier it was stated that some insecticides are soluble in neither oil nor water, but are soluble in one of the exotic solvents, making the formulation quite expensive. To handle the problem, the technical material is blended with one of the dust diluents and a small quantity of water, leaving the insecticide-diluent mixture finely ground but wet. This "wet blend" mixes well with water and can be sprayed with the same tank-settling characteristic as a wettable powder.

3.45 Flowables differ from the dusts and wettable powders in that flowables contain ______________.

- - - - - - - - - - - - - - - -

water

Water-soluble powders (SP). Water-soluble powders are self explanatory. Here, the technical grade material is a finely ground or pelletized water-soluble solid and occasionally contains a small quantity of wetting agent to assist its solution in water. It is added to the spray tank where it immediately dissolves. Unlike the wettable powders and flowables, these formulations do not require constant agitation, since they are true solutions and form no precipitate.

3.46 Soluble powders form ______________ in water and do not require ______________.

- - - - - - - - - - - - - - - -

true solutions agitation

Ultra-low-volume concentrates (ULV). Ultra-low-volume concentrates are normally the technical product in either its original liquid form or its solid form (the original product dissolved in a minimum of solvent). They are frequently applied without further dilution as an extremely fine spray by special aerial or ground spray equipment that limits the volume from one-half pint to a maximum of one-half gallon per acre. ULV formulations are used where good insect control can be obtained, which allows economizing through the elimination of high spray volumes varying from 3 to 10 gallons per acre. This technique has proved quite useful when insect control is desired over vast areas. Because of the necessary fineness of the spray droplets, however, drift is an inherent problem.

3.47 The maximum volume of ULV spray is ______________ per acre.

- - - - - - - - - - - - - - - -

one-half gallon

Dusts

Historically dusts (D) have been the simplest formulations of insecticides to manufacture and the easiest to apply. An example of the *undiluted toxic agent* is sulfur dust used for mite or plant-bug control. An example of the *toxic agent with an active diluent* would be one of the insecticide dusts having sulfur as its *carrier, or diluent*. A *toxic agent with an inert diluent* is the most common type of dust formulation in agricultural use today.

3.48 Commonly used dust formulations have ___________ diluents. A synonym for diluent is ______________.

- - - - - - - - - - - - - - - -

inert carrier

Irrespective of the ease with which they are handled, formulated, and applied, dusts are the least effective, and consequently the least economical, of the insecticide formulations: They give very poor deposit on the target plants. For instance, an aerial application of a standard dust formulation of insecticide will result in 10 to 40 percent of the material reaching the crop. The remainder drifts upward and downwind. Dusts are thus not the formulation of choice in IPM. Psychologically, dusts are annoying to the nongrower who sees great clouds of insecticide resulting from an aerial application; in contrast, the grower believes he is receiving a thorough application for the very same reason. Under similar circumstances, an aerial application of a water emulsion spray deposits 50 to 80 percent of the insecticide on target.

3.49 Sprays are the formulation of choice in IPM situations. Why?

\- - - - - - - - - - - - - - - - -

Sprays leave much higher deposits on target than do dusts.

Granular Insecticides

Granular insecticides (G) overcome the disadvantages of dusts in their handling characteristics. Granules are small pellets formed from various inert clays impregnated with toxicant to give the desired content. They range in size from 20 to 80 mesh, the number of grids per inch of screen through which they will pass. The active ingredient of granules ranges from 2 to 25 percent. Granules, which are used almost exclusively in agriculture, can be applied at virtually any time of day, since they can be released aerially in winds of up to 20 mph without problems of drift, a task impossible with sprays or dusts. They also lend themselves to soil application in the drill at planting time to protect roots from insects or to introduce a systemic to the roots for transport to aboveground parts.

3.50 Mesh, when used in connection with granulars, refers to ______________________________.

\- - - - - - - - - - - - - - - - -

screen grid divisions per inch

Granular insecticides are probably the most useful insecticide formulation for IPM in that they kill fewer beneficial insects, including foraging honeybees, when applied broadcast to crops. Their weight prevents lodging on plant surfaces, thus leaving almost no residues for beneficial insects to contact.

3.51 Theoretically, how many 20-mesh granules would fit into a one-cubic-inch container? ______________ 40-mesh? ______________

\- - - - - - - - - - - - - - - - -

- - - - - - - - - - - - - - - - -

8000 64000

3.52 One outstanding advantage of granules over other formulations is that they can be applied __________________________.

- - - - - - - - - - - - - - - - -

at any time

Fertilizer-insecticide Combinations

Fertilizer-insecticide combinations or mixtures (FM) have been made available on special order to growers for years. Fertilizer and insecticide can then be applied in one operation to the soil during planting.

Encapsulated Insecticides

Encapsulated insecticides are new and may find unique uses in agriculture. The insecticide is wrapped in very tiny cylinders of thin polyvinyl plastic, which release the toxicant at varying rates, resulting in a delayed or "slow-release" product.

This formulation can probably be better understood when compared with one of the slow-release cold medicines advertised on television. Encapsulated materials could be used in mosquito abatement programs, where one aerial application of larvacide to stagnant waters might last an entire mosquito breeding season. Encapsulation increases the effective life of a highly volatile insecticide on plants from minutes to days.

3.53 Two benefits from encapsulating an insecticide are that it ______________________________________ and ____.

- - - - - - - - - - - - - - - - -

reduces the number of applications
increases the life of highly volatile material

The remainder of the formulations listed in Table 3.3 are self explanatory and make up but a very small portion of the total insecticides used.

In summary, insecticides are formulated to improve their safety, storage, handling, application, and effectiveness.

3.54 Name some of the potential problems resulting from using nonformulated, technical grade insecticides.

- - - - - - - - - - - - - - - - - -

Great hazard to the applicator
Dangerous to bystanders, domestic animals, and pets
Contaminate food and water supplies
Excessive cost
Difficult to apply
Prolonged residual life
Storage qualities and shelf life likely to be short
Toxic to plants (phytotoxic)
Reduced effectiveness

References

Van Valkenburg, W. 1973. Pesticide formulations. Marcel Dekker, New York. 481 pp.
Ware, G. W. 1975. Pesticides: An autotutorial approach. W. H. Freeman and Company, San Francisco. 205 pp.

Insecticide Nomenclature, Classification, and Modes of Action

Knowledge of insecticides involves among other things, learning about their structure and their name, or nomenclature. For example, let us use Furadan for illustration:

(I) CARBOFURAN (II) (Furadan®)

(III)

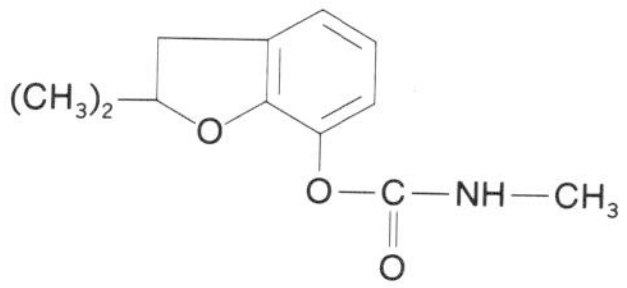

(IV) 2,3-dihydro-2,2-dimethyl-7-benzofuranyl methylcarbamate

The name at the top left, CARBOFURAN (I), is the common name for the compound. Common names are selected officially by the appropriate professional scientific society and approved by the American National Standards Institute (formerly United States of America Standards Institute) and the International Organization for Standardization. Common names of insecticides are selected by the Entomological Society of America, of herbicides by the Weed Science Society of America, and of fungicides by the American Phytopathological Society.

The proprietary name, trade name, or brand name (Furadan®) (II) for the insecticide is given by the manufacturer or by the formulator. It is not uncommon to find six or more brand or trademark names given to a particular insecticide by various formulators. To illustrate, Furadan® is also known as Curaterr®, Bay 70143, and FMC 10242. The latter two are code numbers assigned to the compound by the basic manufacturer when it was first synthesized in the laboratory. Common names are assigned to avoid the confusion resulting from the use of several trade names, as just illustrated. The structural formula (III) is the printed picture of the insecticide molecule. The long chemical name (IV) beneath the structural formula is just that, the chemical name. It is usually presented according to the principles of nomenclature used in Chemical Abstracts, a scientific abstracting journal which is generally accepted as the world standard for chemical names.

3.55 Why are common names assigned to most insecticides?

- - - - - - - - - - - - - - - - - -

To avoid confusion caused by using its numerous trade names.

3.56 In this example, assign the appropriate titles to the blanks:

(I) CARBOPHENOTHION

(II) (Trithion®)

(III) 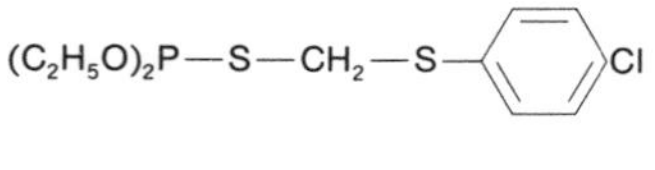

(IV) S-[(*p*-chlorophenylthio) methyl]*O*,*O*-diethyl phosphorodithioate

trade name_____; chemical name _____;
common name_____; structural formula_____.

- - - - - - - - - - - - - - - - - -

- - - - - - - - - - - - - - - -

trade name II; chemical name IV;
common name I; structural formula III.

Insecticides are usually classified by the route they enter the insect: Contact poisons, which penetrate through the outer covering or cuticle of the insect; stomach poisons, which must be eaten by the insect to be effective; and fumigants, which enter through the breathing portals, the spiracles. This is an archaic method of insecticide classification, but it does serve to illustrate the general routes of insect penetration. Under the right circumstances a single insecticide could enter by all three routes.

3.57 The old method of classification divides insecticides into ________________, ____________________, and ____________________ poisons.

- - - - - - - - - - - - - - - -

contact stomach fumigant

A more precise classification is based on the chemical constituents of the insecticide. For the sake of brevity, we will confine our discussion to only four groups of insecticides, because 97 percent of the materials used in agriculture today fall in these categories. They are the organochlorines, the organophosphates, the carbamates, and the formamidines.

3.58 Organochlorines, organophosphates, carbamates, and the formamidines make up _________ percent of the insecticides used on agricultural crops.

- - - - - - - - - - - - - - - -

97

Organochlorines

The organochlorines are insecticides that contain carbon, chlorine, hydrogen, and oxygen. They are also referred to by such names as chlorinated insecticides, chlorinated hydrocarbons, chlorinated organics, and chlorinated synthetics.

DDT and relatives. DDT was placed under federal ban on January 1, 1973, by the EPA in a statement declaring it to be an environmental hazard owing to its long residual life and accumulation, along with its metabolite DDE, in food chains where it proved to be detrimental to certain forms of wildlife. Chemically, DDT belongs to the diphenyl aliphatics, as do TDE (DDD), methoxychlor, dicofol, chlorobenzilate, and Perthane®. Dicofol and chlorobenzilate have no insecticidal qualities but are very effective acaricides.

The way the diphenyl aliphatics kill (the biochemical lesion, or *mode of action*) is not clearly understood. They affect neurons in ways that prevent normal transmission of nerve impulses in insects and mammals, eventually causing the nerve cells (neurons) to fire spontaneously, in turn causing the muscles to twitch. As none of the proposed theories have been clearly proved, it is sufficient to state that in a complex way the diphenyl aliphatics destroy the delicate balance of sodium and potassium within the neurons, preventing them from conducting impulses normally.

DDT

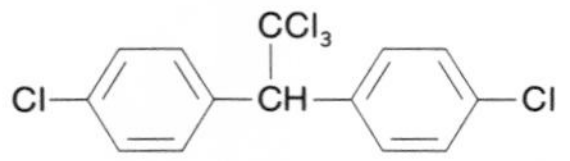

1,1,1-trichloro-2,2-bis(*p*-chlorophenyl)ethane

TDE (DDD)

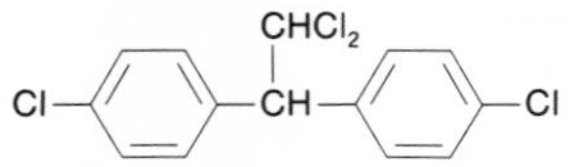

1,1-dichloro-2,2-bis(*p*-chlorophenyl)ethane

METHOXYCHLOR

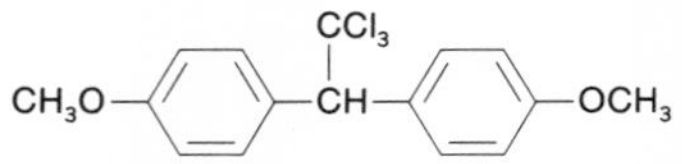

1,1,1-trichloro-2,2-bis(*p*-methoxyphenyl)ethane

DICOFOL

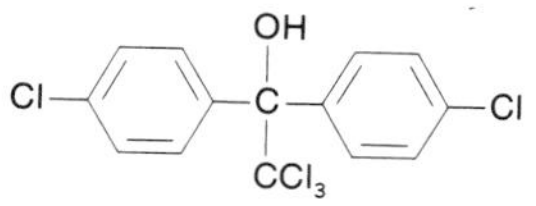

4,4'-dichloro-α(trichloromethyl)benzhydrol

CHLOROBENZILATE

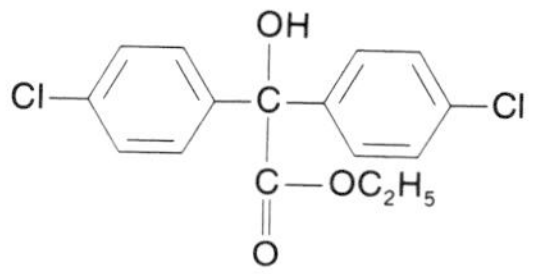

ethyl 4,4'-dichlorobenzilate

Perthane®

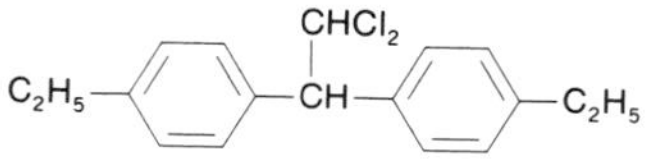

1,1-dichloro-2,2-bis(*p*-ethylphenyl)ethane

DDT and TDE are the only diphenyl aliphatics that have chemical stability or *persistence*. Persistence, as used here, implies a chemical stability giving the insecticides long lives in soil and aquatic environments, and in animal and plant tissues. They are not readily broken down by microorganisms, enzymes, heat, or ultraviolet light. From the insecticidal viewpoint these are good characteristics; from the environmental viewpoint, they are not. In relation to these qualities, the remaining DDT relatives are nonpersistent.

3.59 The only diphenyl aliphatics in the class of organochlorine insecticides considered persistent are ___________ and ___________.

- - - - - - - - - - - - - - -

DDT TDE

3.60 In your own words what are the characteristics of a persistent insecticide?

- - - - - - - - - - - - - - - - -

chemical stability; long life in soil, aquatic enironment, plant and animal tissues; not readily broken down by microorganisms, enzymes, heat or ultraviolet light

---

Cyclodienes. Cyclodienes, another chemically persistent group of insecticides, have proved their usefulness primarily as soil insecticides, though they were applied to the foliage of many crops in the 1950's and 1960's. The cyclodienes are of more recent origin than DDT (1939). The following eight materials were first described or patented in the year indicated: chlordane, 1945; aldrin and dieldrin, 1948; heptachlor, 1949; endrin, 1951; mirex, 1954; endosulfan, 1956; and Kepone, 1958. There are others, but they are of minor significance.

The cyclodienes are persistent insecticides in soil and are relatively stable to the action of sunlight. As a consequence they have been used in greatest quantity as soil insecticides for the control of soil-borne insects whose larval stages feed on plant roots. Owing to excessive residues, however, their persistent quality has resulted in their decline as foliage insecticides. Further decline has resulted from soil-insect resistance. Chlordane, aldrin and dieldrin for example offered protection from subterranean termites in treated structures for more than twenty-seven years, indicating their resistance to chemical degradation. However, several soil insects have become resistant to these materials in agriculture resulting in a further decline of their use. This was not true, however, for the wireworm, grub and cutworm complex. Because of the long-range plans of the EPA to phase out the persistent insecticides, most of the cyclodienes will have all agricultural registrations removed by 1976. The cancelation of all agricultural uses for aldrin and dieldrin by the EPA in mid-1975 was considered to be a real loss to certain phases of agriculture. A further consequence was Shell Chemical Company's immediate decision to terminate manufacture and sale of these two insecticides in the United States. Agricultural uses for heptachlor and chlordane were scheduled for cancelation by the EPA in late 1975.

3.61 An important characteristic of the cyclodienes is their ____________, which makes them ideal for the control of ______________ insects.

- - - - - - - - - - - - - - - - -

- - - - - - - - - - - - - - - -

persistence, or stability soil

The mode of action of this class of organochlorine insecticides also is not known. They are neurotoxicants, and have effects similar to those of DDT. They appear to affect all animals in generally the same way, first with nervous activity followed by tremors, convulsions, and collapse. Undoubtedly the cyclodienes also disrupt the delicate sodium-potassium balance of the neurons, but in ways differing slightly from that of DDT.

3.62 The chlorinated cyclodienes act as ______________, whose action is the disruption in the balance of ______________ and ______________ in the neurons.

- - - - - - - - - - - - - - - -

neurotoxicants sodium potassium

DIELDRIN

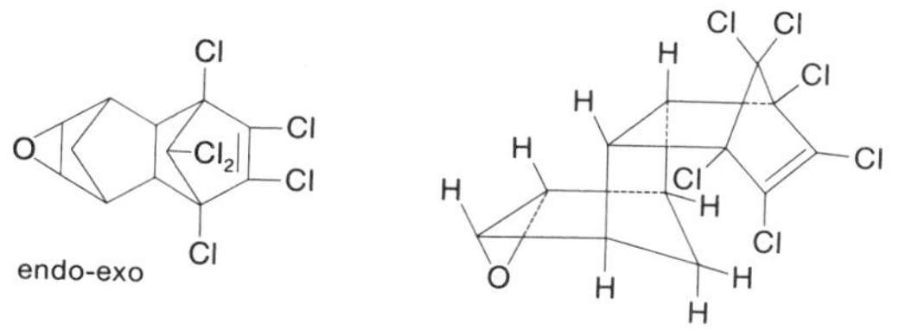

1,2,3,4,10,10-hexachloro-6,7-epoxy-1,4,4*a*,5,6,7,8,8*a*-octahydro-1,4-*endo-exo*-5,8-dimethanonaphthalene

ENDRIN

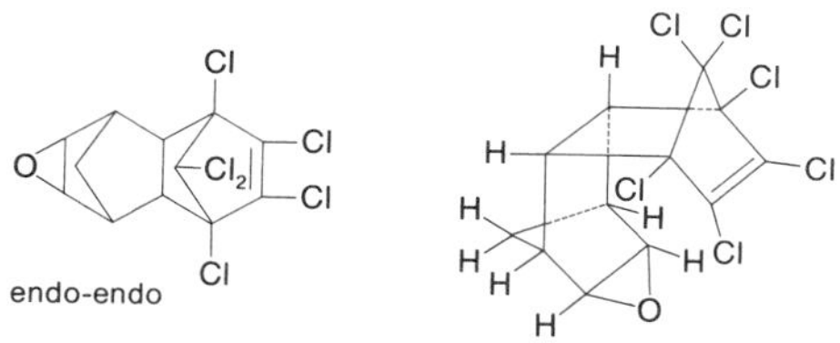

1,2,3,4,10,10-hexachloro-6,7-epoxy-1,4,4*a*,5,6,7,8,8*a*-octahydro-1,4-*endo-endo*-5,8-dimethanonaphthalene

CHLORDANE

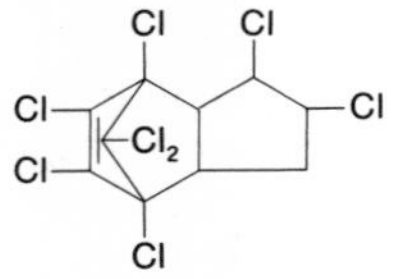

1,2,4,5,6,7,8,8-octachloro-3*a*,4,7,7*a*-tetrahydro-4,7-methanoindane

HEPTACHLOR

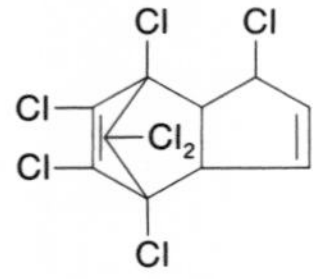

1,4,5,6,7,8,8-heptachloro-3*a*,4,7,7*a*-tetrahydro-4,7-methanoindene

ALDRIN

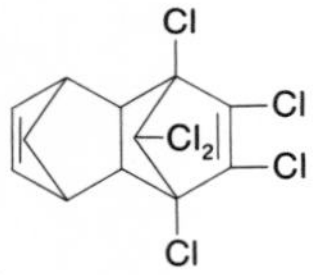

1,2,3,4,10,10-hexachloro-1,4,4*a*,5,8,8*a*-hexahydro-1,4-*endo*-*exo*-5,8-dimethanonaphthalene

ENDOSULFAN (Thiodan®)

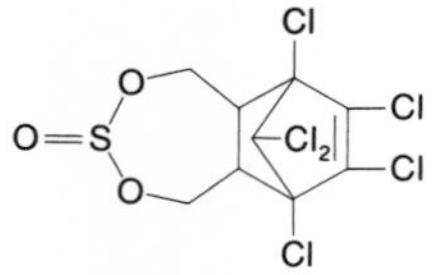

6,5,8,9,10,10-hexachloro-1,5,5*a*,6,9,9*a*-hexahydro-6,9-methano-2,4,3-benzodioxathiepin 3-oxide

MIREX

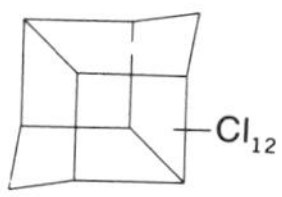

dodecachlorooctahydro-1,3,4-metheno-1*H*-cyclobuta[*cd*]pentalene

KEPONE

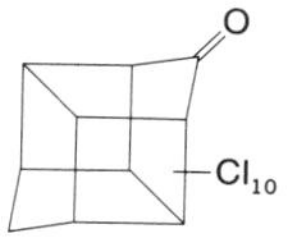

decachlorooctahydro-1,3,4-metheno-2*H*-cyclobuta[*cd*]pentalen-2-one

The chlorinated cyclodienes are generally equitoxic; they usually have equal toxicity to insects, mammals, birds, and fish. That is, given the same dosage based on weight, such as milligrams per kilogram of body weight, these materials have about the same degree of toxicity. There are always exceptions, and fish are more susceptible since they are totally surrounded when the compound is introduced into water. They literally eat, sleep, and breathe the toxicant in their aquatic environment.

3.63 Insecticides having equal toxicity to different species of animals on a weight/weight basis are said to be ________________ to these animals.

- - - - - - - - - - - - - - - -

equitoxic

3.64 Fish are particularly sensitive to the toxic effects of the organochlorine insecticides. Why?

- - - - - - - - - - - - - - - -

Fish eat, breathe, and are surrounded by the insecticides, which are chemically very stable.

<u>Polychloroterpenes</u>. There are only two very similar materials in the polychloroterpene group: toxaphene, discovered in 1947; and strobane, introduced in 1951. Toxaphene has by far the greater use, which is almost exclusively on cotton. It has rather low toxicity to insects

alone and is thus formulated with other insecticides, usually methyl parathion, a member of the organophosphates. Both toxaphene and strobane have modes of action similar to the cyclodiene insecticides and are sometimes classed with the cyclodienes. Toxaphene, second only to DDT, is the most economical of the synthetic organic insecticides. Its average wholesale quotation per pound for the first half of June 1973 was twenty-five cents. DDT during that same interval was quoted at twenty-two cents per pound. It is this cost factor that is influential in the tremendous amount of toxaphene used on cotton, despite its nominal insecticidal qualities. In combination with DDT at the rate of 2:1, it was found to be quite effective against boll weevils, whereas either product alone was essentially ineffective.

3.65 Why were toxaphene and DDT used in such great quantity on cotton prior to 1973?

- - - - - - - - - - - - - - - -

They were economical and effective.

It was the toxaphene-DDT combination that retained toxaphene's usefulness to agriculture, which later carried over to other combinations when all registered uses for DDT were canceled by EPA in 1973. Toxaphene is classed as a semipersistent material in that it does persist in soil and water environments but does not persist in animal tissue for any significant length of time, and does not build up in food chains as do DDT, DDE, and dieldrin, for example.

TOXAPHENE

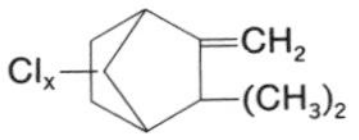

$C_{10}H_{10}Cl_8$

chlorinated camphene containing 67-69% chlorine

STROBANE

$C_{10}H_{11}Cl_7$

terpene polychlorinates (65% chlorine)

3.66 Toxaphene and strobane are considered ________ persistent insecticides.

- - - - - - - - - - - - - - - -

semi

Organophosphates

The organophosphates (OPs) are also known as *organic phosphates, phosphate insecticides, phosphorus insecticides, nerve gas relatives, phosphates,* and *phosphorus esters* (or *phosphoric acid esters*). They are derivatives of phosphoric acid and are generally the most toxic class of insecticides to vertebrate and invertebrate animals. Because of their general chemical structure, they are indeed related to the "nerve gases."

Their insecticidal action was discovered during World War II in Germany in the study of materials closely related to the nerve gases, sarin, soman, and tabun. Initially, the discovery was made in the search for a substitute for nicotine, which was in critically short supply.

The OPs have three distinctions: They are generally much more toxic to humans and other vertebrates than the organochlorines, they are broader in insecticidal spectrum than the organochlorines, and they are nonpersistent. Their nonpersistence brings them onto the agricultural scene as substitutes for the persistent organochlorines, particularly DDT.

3.67 What are the organophosphates' three distinctive credentials?

- - - - - - - - - - - - - - - -

toxic to man and animals
broad spectrum
nonpersistent

As for mode of action, the OPs exert their toxic action by tying up or inhibiting certain important enzymes of the nervous system, in both vertebrates and invertebrates. These enzymes are cholinesterases (ChE). Throughout the nervous system there are electrical switching centers, or synapses, where the electrical signal is carried across gaps between neuron and muscle, or between neuron and neuron. These gaps are crossed by a conducting chemical—acetylcholine (ACh) in most instances. After the electrical signal (nerve impulse) has been transmitted across the gap by ACh, the ChE enzyme moves in quickly and removes the ACh to avoid a "jammed" circuit. These chemical reactions happen extremely rapidly and go on constantly under normal conditions. When OPs enter an animal, they attach to the ChE in a way that prevents the ChE from removing the ACh, resulting in jammed circuits owing to the ACh accumulation. In other words, accumulation of ACh interferes with the neuromuscular junction, giving rise to rapid twitching of voluntary muscles and finally to paralysis. This is of great importance in proper functioning of the respiratory system. The mode of action of the organophosphates is not quite as simple as outlined here, but such a description should serve the purposes of this generalized study.

3.68 An organophosphate insecticide, abbreviated to __________, acts by inhibiting enzymes known as __________________________, abbreviated to _______.

- - - - - - - - - - - - - - - - -

OP cholinesterases ChE

The organophosphate insecticides are further divided into three classes: the aliphatic, the phenyl, and the heterocyclic derivatives. Each class contains several distinctive compounds.

Aliphatic derivatives. Aliphatic literally means "carbon chain," and the linear arrangement of carbon atoms differentiates them from the ring or cyclic structures. All of the aliphatic OPs are simple phosphoric acid derivatives linked to short carbon chains. Among these are malathion, a safe insecticide for home and pets; trichlorfon, used on crops and for fly control on the farm; and monocrotophos, an OP with plant systemic qualities and a high mammalian toxicity.

MALATHION

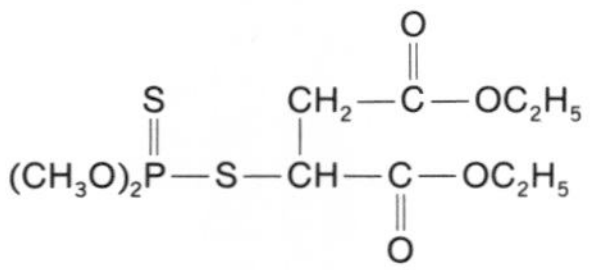

diethyl mercaptosuccinate, *S*-ester with *O,O*-dimethyl phosphorodithioate

TRICHLORFON (Dylox®)

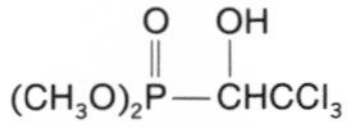

dimethyl (2,2,2-trichloro-1-hydroxyethyl)phosphonate

MONOCROTOPHOS (Azodrin®)

$$(CH_3O)_2\overset{\overset{O}{\|}}{P}-O-\underset{\underset{CH_3}{|}}{C}=CH\overset{\overset{O}{\|}}{C}-NH-CH_3$$

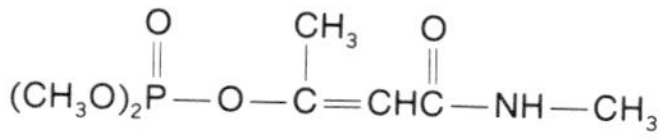

3-hydroxy-N-methyl-*cis*-crotonamide dimethyl phosphate

Plant systemic insecticides are those taken into the roots and translocated to the aboveground parts, where they are toxic to any sucking insects feeding on plant juices. Normally caterpillars and other plant-tissue-feeding insects are not controlled since they do not ingest enough of the systemic-containing juices to be affected.

3.69 Why are systemic insecticides usually ineffective against insects with chewing mouthparts?

They do not eat enough of the systemic to be effective.

Within the aliphatic derivatives are several plant systemics: dimethoate, dicrotophos, oxydemetonmethyl, disulfoton, demeton, and phorate.

Mevinphos is a highly toxic OP used in vegetable production because of its very short insecticidal life. It can be applied up to one day before harvest for insect control and leave no residues on the crop after harvest.

MEVINPHOS (Phosdrin®)

$$(CH_3O)_2\overset{\overset{O}{\|}}{P}-O-\underset{\underset{CH_3}{|}}{C}=CH\overset{\overset{O}{\|}}{C}-OCH_3$$

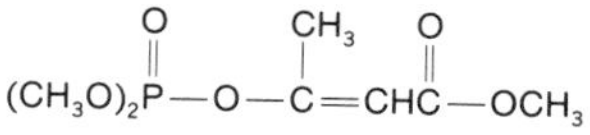

methyl 3-hydroxy-*alpha*-crotonate, dimethyl phosphate

The aliphatic organophosphate insecticides are the simplest in structure of the organophosphate molecules. They have a wide range of toxicities, and several possess a relatively high water solubility, giving them plant systemic qualities.

DIMETHOATE (Cygon®)

$$(CH_3O)_2\overset{\overset{S}{\|}}{P}-S-CH_2\overset{\overset{O}{\|}}{C}-NH-CH_3$$

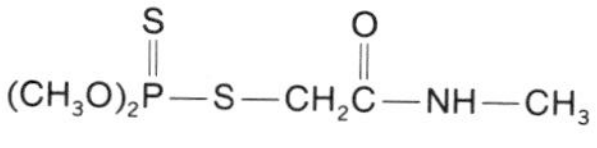

O,O-dimethyl S-(N-methylcarbamoylmethyl) phosphorodithioate

DISULFOTON (Di-Syston®)

$$(C_2H_5O)_2\overset{\overset{S}{\|}}{P}-S-CH_2CH_2-S-C_2H_5$$

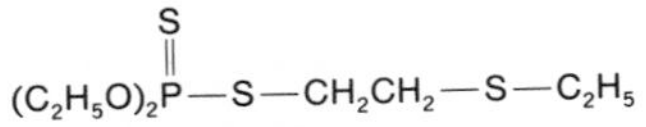

O,O-diethyl S-2-[(ethylthio)ethyl]phosphorodithioate

DEMETON (Systox®)

$$(C_2H_5O)_2\overset{\overset{O}{\|}}{P}-S-CH_2CH_2-S-C_2H_5$$

$$(C_2H_5O)_2\overset{\overset{S}{\|}}{P}-O-CH_2CH_2-S-C_2H_5$$

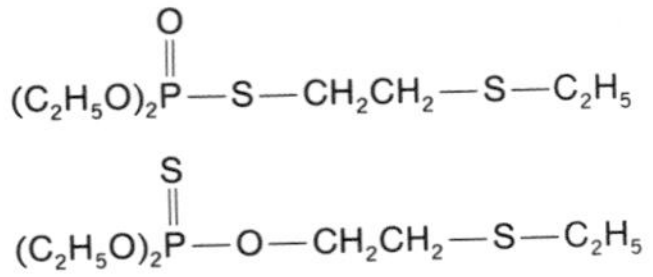

mixture of O,O-diethyl S-(and O)-2-[(ethylthio)ethyl]phosphorothioates

PHORATE (Thimet®)

$$(C_2H_5O)_2\overset{\overset{S}{\|}}{P}-S-CH_2-S-C_2H_5$$

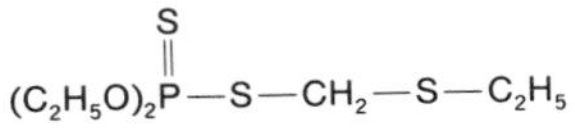

O,O-diethyl-S-[(ethylthio)methyl] phosphorodithioate

Phenyl derivatives. The phenyl OPs contain a benzene ring with one of the ring hydrogens displaced by attachment to the phosphorus moiety and others frequently displaced by Cl, NO_2, CH_3, CN, S, and so on. The phenyl OPs are generally more stable than the aliphatic OPs, resulting in longer lasting residues.

Ethyl parathion is the most familiar of the phenyl OPs, being the second phosphate insecticide introduced into agriculture (1947). The first, TEPP, was introduced in 1946. Because of its age and utility, ethyl parathion's usage is greater than that of many of the less useful materials combined.

ETHYL PARATHION

$$(C_2H_5O)_2\overset{\overset{S}{\|}}{P}-O-\underset{}{\bigcirc}-NO_2$$

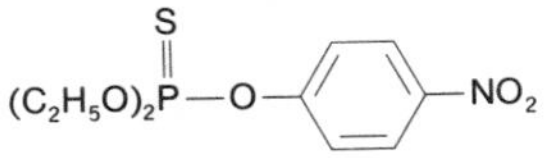

O,O-diethyl O-p-nitrophenyl phosphorothioate

Methyl parathion became available in 1949 and proved to be more useful than ethyl parathion because of its lower toxicity to man and domestic animals and its broader range of insect control. It has a shorter residual life, making it more desirable in certain instances than ethyl parathion.

METHYL PARATHION

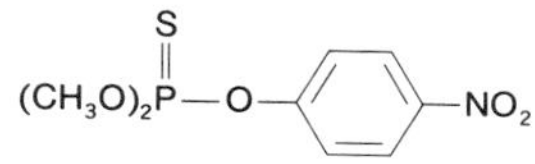

O,O-dimethyl *O-p*-nitrophenyl phosphorothioate

Several systemic insecticides are also found in the phenyl OPs and are usually animal systemics; ronnel and crufomate are examples.

RONNEL (Korlan®)

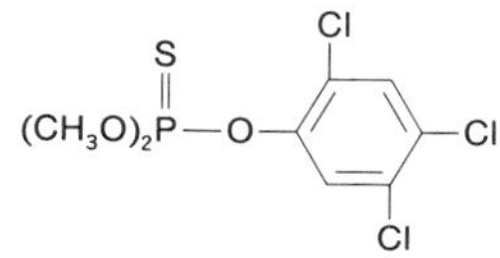

O,O-dimethyl *O*-2,4,5-trichlorophenyl phosphorothioate

CRUFOMATE (Ruelene®)

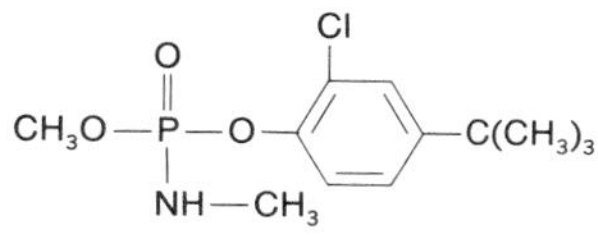

4-*tert*-butyl-2-chlorophenyl methyl methylphosphoramidate

Another home-safe OP, much like malathion in its overall usefulness against home and garden pests, is Gardona .

Gardona®

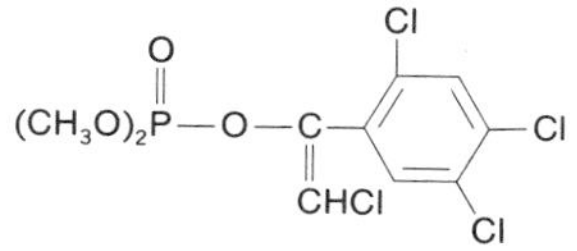

2-chloro-1-(2,4,5-trichlorophenyl)vinyldimethyl phosphate

<u>Heterocyclic derivatives</u>. The term heterocyclic indicates that the ring structures are composed of unlike atoms. In a heterocyclic carbon compound, for example, one or more of the carbon atoms is displaced by oxygen, nitrogen, or sulfur, and the ring may have three, five, or six atoms.

3.70 Heterocyclic indicates that cyclic molecules are made of ________________ atoms.

- - - - - - - - - - - - - - - -

unlike

The first insecticide made available in this group is probably diazinon (1952). Note that the six-membered ring contains two nitrogen atoms, very likely the source of its proprietary name, since one of the constituents used in its manufacture is pyrimidine, a diazine.

DIAZINON

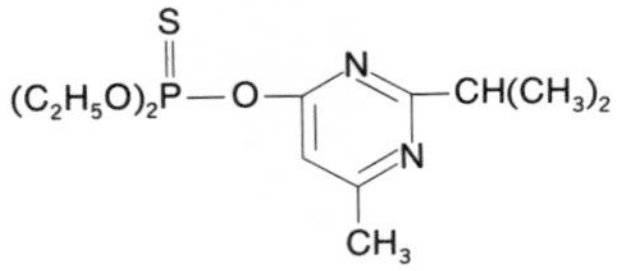

O,O-diethyl *O*-(2-isopropyl-4-methyl-6-pyrimidyl) phosphorothioate

Azinphosmethyl, the second oldest member of this group used in U. S. agriculture (1954), serves as both an insecticide and an acaricide on cotton.

AZINPHOSMETHYL (Guthion®)

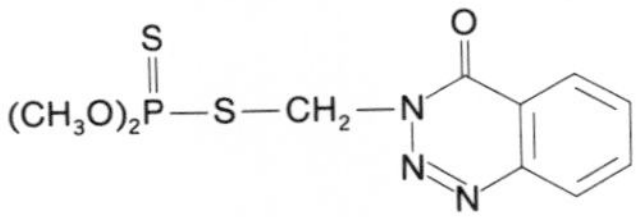

O,O-dimethyl *S*(4-oxo-1, ,3-benzotriazin-3(4*H*)-ylmethyl) phosphorodithioate

The heterocyclic OPs are complex molecules and generally have longer lasting residues than many of the aliphatic or phenyl derivatives. Their breakdown products (metabolites) are frequently many, making their residues sometimes difficult to measure in the laboratory. Consequently, their use on food crops is somewhat less than either of the other two groups of phosphorus-containing insecticides.

3.71 Are there any basic differences between the mode of action of the aliphatic, phenyl, and heterocyclic organophosphates?

- - - - - - - - - - - - - - - -

- - - - - - - - - - - - - - - -

No.

Carbamates

Carbamate insecticides have a mode of action similar to that of the organophosphates: the inhibition of the vital enzyme cholinesterase (ChE). The carbamates were introduced in 1951, but fell by the way because they were ineffective. Those early attempts concentrated on the N-dimethyl carbamates, generally less toxic than the N-methyl carbamates which comprise the bulk of the currently used materials.

3.72 The carbamate insecticides act by inhibiting ____________________.

- - - - - - - - - - - - - - - -

ChE

Carbaryl, introduced in 1956, was the first successful carbamate. More of it has been used the world over than all the remaining carbamates combined. Two qualities have elevated it to the popularity that it currently receives: very low mammalian oral and dermal toxicity, and a rather broad spectrum of insect control. As you can see, carbaryl is an N-methyl compound.

CARBARYL (Sevin®)

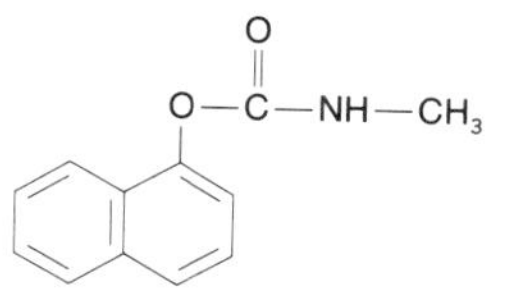

1-naphthyl methylcarbamate

One carbamate more recently developed is methomyl, which has been extremely useful, especially for worm control on several crops.

METHOMYL (Lannate®, Nudrin®)

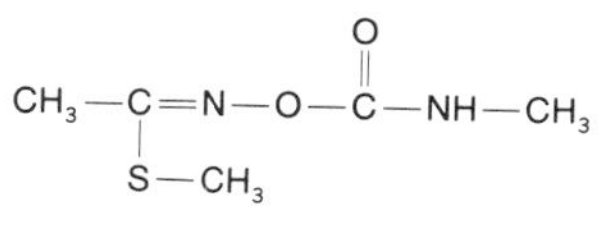

methyl *N*-[(methylcarbamoyl)oxy]thioacetimidate

Some of the carbamates are plant systemics, indicating that they have a rather high water solubility in order to be taken into the roots or leaves. They are also not readily metabolized by the plants. Methomyl, aldicarb, and carbofuran are carbamates that have distinct systemic characteristics.

ALDICARB (Temik®)

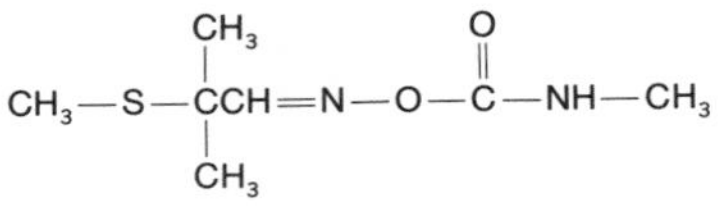

2-methyl-2-(methylthio) propionaldehyde-*O*-(methylcarbamoyl) oxime

CARBOFURAN (Furadan®)

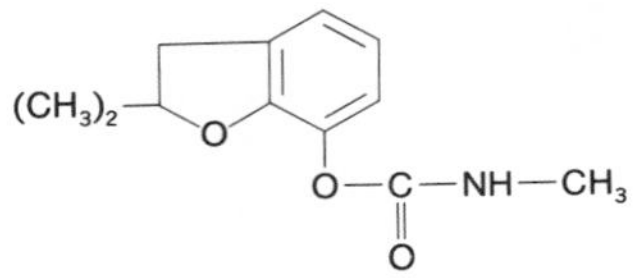

2,3-dihydro-2,2-dimethyl-7-benzofuranyl methylcarbamate

Other carbamates of interest are given below.

MEXACARBATE (Zectran®)

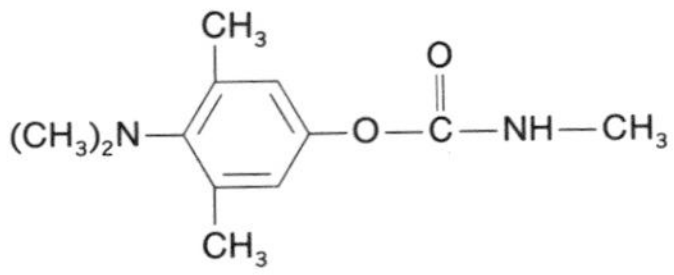

4-dimethylamino-3,5-xylyl methylcarbamate

FORMETANATE (Carzol®)

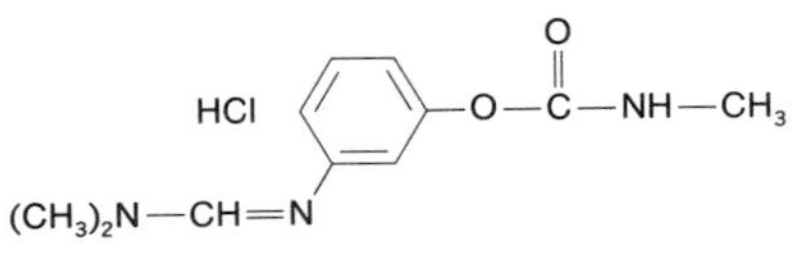

N,*N*-dimethyl-*N*'[3-[[(methylamino)carbonyl]=
oxy]phenyl]methanimidamide

In summary, the carbamates are inhibitors of cholinesterase, are plant systemics in several instances, and are for the most part somewhat narrower in spectrum than the organophosphates.

Formamidines

The formamidines offer a relatively new class of insecticides which are both ovicidal (kill eggs) and larvicidal. Two examples are chlordimeform and U-36059. They are effective against the eggs and very young caterpillars of several moths of agricultural importance and are also effective against most stages of mites and ticks.

CHLORDIMEFORM (Galecron®, Fundal®)

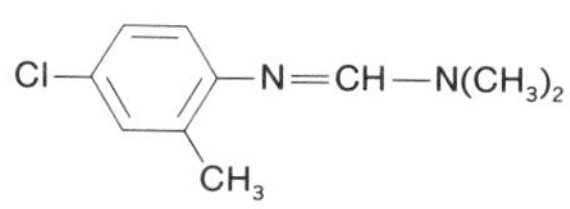

N'-(4-chloro-*o*-tolyl)-*N*,*N*-dimethylformamidine

U-36059

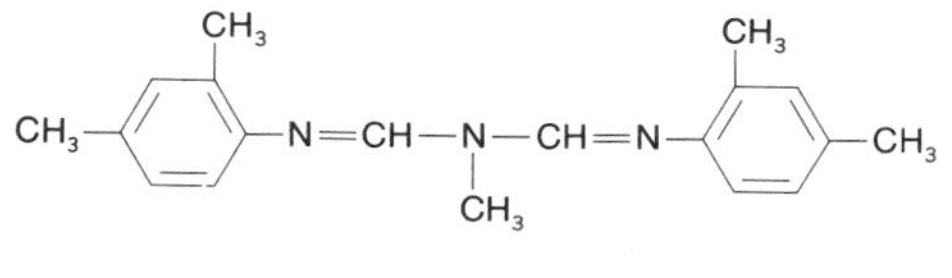

N-methyl-*N*'-2,4-xylyl-*N*-(*N*-2,4-xylylformimidoyl)formamidine

Formamidines are currently most useful in the control of organophosphate- and carbamate-resistant pests. Poisoning symptoms are distinctly different from other materials. It has been proposed that one possible mode of action is the inhibition of a previously unmentioned enzyme, *monoamine oxidase*. This results in the accumulation of compounds termed *biogenic amines*, whose actions are not fully understood. However, they may act in certain instances as chemical transmitters of synapses, similar to acetylcholine. Thus, the formamidines introduce a new mode of action for the insecticides and acaricides. It is predicted that many of the new compounds of the future will belong to the formamidines.

3.73 The formamidines present a new mode of action among the insecticides, namely, the inhibition of ______________________.

- - - - - - - - - - - - - - -

monoamine oxidase

Factors Affecting Efficacy

Several factors are important in determining the effectiveness of insecticide applications. Giving attention to these can greatly improve the efficiency of insecticides as one of the tools of pest management.

Choice of the proper insecticide is one factor which affects efficacy in a given situation. Insecticides are not equally effective in insect control. For example, the insecticide azinphosmethyl gives excellent control of the pink bollworm in cotton but is poor in controlling the cotton bollworm. Methyl parathion, on the other hand, gives good control of the pink bollworm, although somewhat less effective control than azinphosmethyl, and good control of the cotton bollworm. From the standpoint of effectiveness, azinphosmethyl would be the proper choice if pink bollworm were the only pest of concern, whereas methyl parathion would be a better choice for controlling a combination of the two pests or an infestation of the cotton bollworm. Properly identifying the pest or pest complex is necessary if insecticide effectiveness and efficiency are to be maximized.

Properly timed insecticide applications result in the best possible insect control. Timing is important from the standpoint of when an application is made relative to time of day and to the amount of insect infestation, and to the stage of insect development.

Insecticides are most effective when applied in relatively calm weather conditions. Airplane applications are usually most effective when made early and late during the day or at night. More insecticide reaches the target area at these times because of less evaporation and drift. Applications should generally be avoided during the time period between 10:00 a.m. and 4:00 p.m., especially during hot weather. Effectiveness of ground equipment applications is affected less critically by the time of day, but basically the same considerations apply as for air application.

Good timing of insecticide applications is especially important relative to the development of insect infestations. Applications should be made immediately when field sampling indicates the presence of an economic infestation. Follow-up applications should be made at the appropriate interval based on continued field sampling. Practices such as early light-dosage applications to prevent the development of an infestation, extended interval applications when no economic infestation is present, and applications intended to "clean up" fields infested with sub-economic levels of one or more pests should be avoided. These practices often create worse situations than before application, thereby definitely reducing the effectiveness of insecticides used in controlling the pest problems involved and adversely affecting the value of insecticides as a pest-management tool.

3.74 Basically, what is wrong with early light dosages to prevent buildup, extended interval applications when no economic infestation is present, and "clean-up" application with subeconomic infestations?

- - - - - - - - - - - - - - - - - -

These practices create pest problems and are a costly waste of valuable materials.

Thorough coverage in the target area is another important factor in insecticide efficacy. The total gallonage used in applying spray solutions should be adequate to insure good coverage of the crop, with consideration given to the amount of foliage involved and where the insect or insects to be controlled are most active within the plant canopy. For example, in airplane application, 1 gallon of total solution per acre may be adequate for small plants or for pests concentrated in the top of plants, while 3 to 5 gallons per acre may be necessary in the same crop as the plants become large with dense foliage or as pests become concentrated in difficult-to-reach areas of the plant canopy. Usually higher gallonage is used and more thorough coverage obtained by ground application than by air because of the nature of the equipment and the associated attention to nozzle placement relative to plant size, row spacing, and other factors. One of the problems associated with air application is that of missed spots or strips within the target area because of swaths too wide apart, drift caused by wind velocity or direction change, and difficult-to-reach areas under or near power lines, trees and buildings. These untreated or poorly treated areas may serve as reservoirs of certain pests, permitting quicker infestation. Insecticide effectiveness is sometimes questioned when the real problem is poor coverage.

3.75 Generally, which type of insecticide application gives better coverage of plants, ground or air?

- - - - - - - - - - - - - - - - - -

- - - - - - - - - - - - - - - -

Ground.

3.76 Name four factors affecting the efficacy of chemical control.

- - - - - - - - - - - - - - - -

1. insecticide choice
2. proper timing as to hour of day
3. degree of infestation
4. coverage

References

Corbett, J. R. 1974. The biochemical mode of action of pesticides. Academic Press, New York. 330 pp.

O'Brien, R. D. 1967. Insecticides action and metabolism. Academic Press, New York. 332 pp.

Spencer, E. Y. 1973. Guide to the chemicals used in crop protection. Publ. 1093, 6th ed., Information Canada, Ottawa, Ontario, Canada. 542 pp.

Ware, G. W. 1975. Pesticides: An autotutorial approach. W. H. Freeman and Company, San Francisco. 205 pp.

Insecticide Application and Drift

Given a choice of application equipment and formulations, which should be used to obtain the least drift off target and the greatest deposit on target? There is no cut-and-dried answer to this question, but considerable data are available to help in decision making.

Let us begin with dusts versus sprays. The facts simply stated are these: (1) Dusts drift more than sprays, and (2) dusts deposit less on target than sprays. Table 3.4 presents convincing information to support the first point. Note that when both are falling 10 feet, medium aircraft spray droplets drift 22 feet in a 3 mph breeze, whereas the usual dust formulations will move 4,436 feet.

3.77 Given a choice of formulations, which would you recommend to give best coverage and, consequently, control, sprays or dusts?

- - - - - - - - - - - - - - - -

Sprays.

Table 3.4 Drift pattern in relation to particle size.*

Drop diameter (microns)	Particle type	Feet particle drifts in 3 MPH wind while falling 10 feet
400	Coarse aircraft spray	8½
150	Medium aircraft spray	22
100	Fine aircraft spray	48
50	Air-carried sprays	178
20	Fine sprays and dusts	1,109
10	Standard dusts and aerosols	4,436
2	Aerosols	110,880

*Modified from, "Principles of Plant and Animal Pest Control," Vol. 3, Insect Pest Management and Control Natl. Acad. Sci. Pub. 1965, 1969.

Figure 3.14 presents composite information on actual field experiments of drift onto distant alfalfa fields from aerially applied sprays and dusts. Sprays deposited

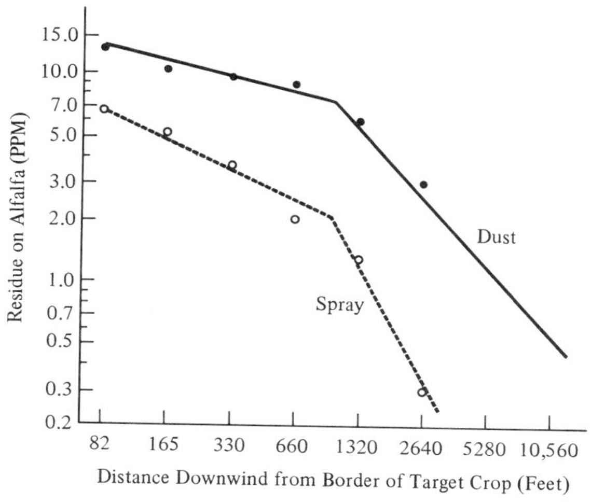

Figure 3.14 Comparison of insecticide dust versus spray drift onto alfalfa.

0.3 ppm residue on alfalfa 2640 feet downwind, whereas the same rates of dust application deposited 3.3 ppm or elevenfold the drift residue.

Regarding the second point above, recent research has shown that aerially applied dusts deposited 14 percent of the application on target, whereas aerially applied sprays averaged 54.5 percent on target.

3.78 Considering the consequences of drift problems to nontarget areas, which should you recommend, sprays or dusts?

- - - - - - - - - - - - - - - -

Again, sprays!

The final comparison remains to be made between ground sprays and aerial sprays. Another recent study has shown that insecticide sprays made to mature cotton plants by a high-clearance ground sprayer, utilizing three nozzles per row, deposited 73 percent on target in short (29 inches) cotton, and 89 percent on target in tall (49 inches) cotton. The average deposit by ground equipment was 82 percent, whereas the average of aerial sprays (above) was 54.5 percent.

3.79 This is similar to an earlier question. Given a choice of ground or aerial applications, all other factors being equal, which would you choose, and why?

- - - - - - - - - - - - - - - -

Ground application, because about 25 percent more of the material reaches the target.

A last comparison of drift from ground and aerial sprays is presented in Figure 3.15. At all distances downwind the aerial applications resulted in four to five times the drift from the high-clearance ground sprayer for both evening and morning treatments.

3.80 Referring to Table 3.4, medium aircraft spray drifts ________ feet in a 3 mph breeze, whereas fine sprays and dusts drift ________ feet.

- - - - - - - - - - - - - - - -

22 feet 1,109 feet.

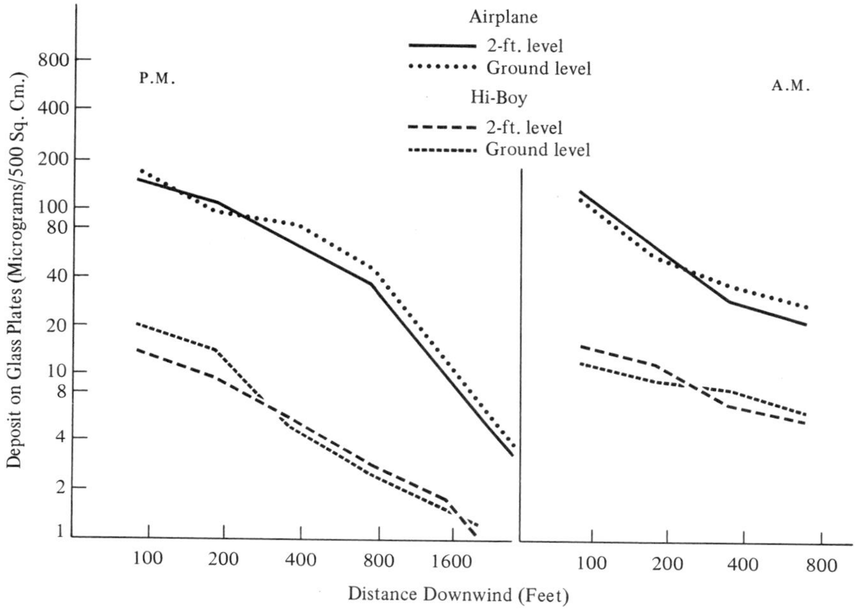

Figure 3.15 Deposits of methoxychlor on glass plates indicating relative drift from evening and morning Hi-Boy and aerial applications. Higley, Arizona, 1967.

In summary: On-target deposits of insecticides from ground sprays > aerial sprays > aerial dusts, and off-target drift of insecticides from aerial dusts > aerial sprays > ground sprays.

References

Ware, G. W., B. J. Estesen, W. P. Cahill, P. D. Gerhardt, and K. R. Frost. 1969. Pesticide drift I. High-clearance vs. aerial application of sprays. J. Econ. Entomol. 62(4):840-843.

Ware, G. W., W. P. Cahill, P. D. Gerhardt, and J. M. Witt. 1970. Pesticide drift IV. On-target deposits from aerial application of insecticides. J. Econ. Entomol. 63(6):1982-1983.

Ware, G. W. 1972. Pesticide drift . . . dust vs. spray. Prog. Agr. Ariz. 24(1):10-11.

Ware, G. W. 1975. Pesticides: An autotutorial approach. W. H. Freeman and Company, San Francisco, 205 pp.

Ware, G. W., W. P. Cahill, and B. J. Estesen. 1975. Pesticide drift VIII. Deposit efficiency from ground sprays on cotton. J. Econ. Entomol. 68:549-550.

Selective Use of Insecticides

An insecticide is a substance used to kill insects. We know, however, that insecticides will also kill many kinds of animals other than insects, although normally they are not used for such purposes. Even within the class of insects, different insecticides affect various insects differently. An insecticide that is deadly against one group may have little, if any, effect on another. Generally, insecticides kill many kinds of insects, but some kill more kinds than others, thus the term *broad-spectrum insecticide*.

3.81 What is a broad-spectrum insecticide?

\- - - - - - - - - - - - - - -

One that kills a wide variety of insects.

The exact meaning of *selective insecticide* is not quite as clear-cut. In one instance it simply means that the chemical is naturally more toxic to one group of insects than to others. To put it another way, most groups of insects are physiologically more tolerant to the chemical than are the few readily killed by it. It can also be used to indicate the differential action between insects and warm-blooded animals.

Another definition of a selective insecticide relates to its being used in such a way that certain insect groups are more adversely affected by it than other groups. From a practical standpoint this approach offers the most immediate hope of being included in a pest-management program. The use of an insecticide to kill a particular pest while causing little effect on other insects, particularly beneficial insects, achieves selective action with the insecticide.

Physiologically, it may be as toxic to the unaffected insects as it is to the treated species, but because of the way it is used it reaches one species and not the others. This may be the result of the timing of the application, the formulation used, the dosage level, or numerous other reasons. The main difference here is that the planning of application takes into account the existence of both the pest and the desirable species, and utilizes some difference in their habits, distribution, or biology that permits the use of the chemical to kill one and not the others. Although some insecticides are inherently more toxic physiologically to certain insects, and some are equally toxic to different insects, they can be used by selective action to affect certain ones more than others. The objective of both types is to use insecticides so that beneficial insects are not destroyed.

3.82 What is a selective insecticide?

- - - - - - - - - - - - - - - - -

It can mean an insecticide that is inherently more toxic to certain insects than to others or one that by virtue of the way it is used, will kill only certain insects while sparing others, that is, one that achieves selective action.

It is safe to say that complete dependence on insecticides for insect control is a thing of the past. It is generally agreed, too, that unless selective insecticides are used, or insecticides are used in a selective manner, the effect of beneficial insects will be minimal in any program requiring considerable insecticide use.

3.83 Use of ______________ insecticides, or insecticides in a selective ________________, is essential to make maximum use of _______________ insects.

- - - - - - - - - - - - - - - - -

selective   manner   beneficial

Various scientists, ranging in interests from toxicology to applied ecology, have stressed the desirability of using both insecticides and beneficial insects in insect-control programs. The toxicologist is generally working toward the development of physiologically selective insecticides, whereas the applied ecologist is seeking ways of using the ones currently available to achieve selective action.

Insect-control strategy

What do we mean by "insect-control strategy"? What strategy has been used and will be used in the future to control our insect pests? Strategy is the general approach to control, including decisions on which method should be used. Using this definition, our strategy for the past twenty-five to thirty years has been complete elimination of the insect pests when possible, and the method used has been chemical control. The prevailing attitude has been that excess dosage levels are desired, which is consistent with the fertilizer philosophy of "If a little does a lot of good, then a lot should . . . ."

If we are to make significant advances in insect control we must think in terms of a new strategy—insect pest management. With this strategy we have the option of using a number of different methods, preferably together in a mutually workable program. In a system such as this, insecticides will continue to be a vital part of the program but will be the stopgap or the method of last resort in keeping the insect pest from economically damaging the crop. In such cases one has to accept the adverse effects of the insecticide on other insects and on the environment along with the benefits to agriculture.

From the practical standpoint, few of our agricultural insecticides are used selectively. In addition to repeated applications, which are highly detrimental to beneficial insects, the dosage levels are generally excessive. Like the adage "Using a sledge hammer to drive a tack," our present use of insecticides is very crude.

3.84 The insect-control strategy of the past thirty years has been total elimination of insect pests through ________________________.

- - - - - - - - - - - - - - -

chemical control

Consideration of economic levels

Assuming that insecticides are overused, and there are plenty of data to support this assumption, what action at the grower level could help prevent this? Of first importance is to use the insecticide only when needed, and this means using economic levels on which to base insecticide treatments. On most agricultural crops there are relatively few key pests, that is, those generally causing economic damage during the growing season.

Economic levels have been discussed earlier and should now be a working part of your IPM concept. Economic levels, of course, vary with different crops and even on the same crop, depending on stage of growth, time of season and the number and kinds of beneficial insects associated with the pest. This is what was meant when we earlier referred to a "dynamic" economic level in Unit 2. Two main problems are evident in our concept of economic levels, both of which must be solved before the growers will generally accept it. First, the pest population level required to cause economic damage must be set—and accepted. This is not easy, and relatively few sound economic injury levels have been established. Unfortunately, in many crops treatment is often made too early or unnecessarily because economic levels are too low, poorly defined, or simply unknown. This should not, however, prevent us from using the best information available to set a safe level on a temporary basis for initiating control. The very fact that sound economic levels are generally lacking has long been used as the basis for full-season insecticide programs. Ecologically, this practice would be unsound even if a truly selective insecticide were available for the pest in question.

The second problem in using economic levels relates to keeping pests below a set level without danger of sustaining losses. If we are to use insecticides in a corrective rather than a preventative manner, then we must be able to predict population trends to some extent. At present, this is not as reliable as would be desired, and so economic levels must be conservative to account for unpredictable population increases. Ability to predict population trends will lead to more conservative use of insecticides, and will make the combination of other control methods with insecticides more feasible.

Evidence of need for selective insecticides

The evidence pointing to the need for selective insecticides is really the same as that discussed earlier which pointed to the need for IPM. And since the chemical control method is one of the tools that can be used in an IPM program, selective insecticides would make this method much more flexible and compatible with other methods, particularly biological control. The use of a selective insecticide would not necessarily eliminate all the adverse effects brought on with broad-spectrum insecticides, but it would eliminate some and reduce others.

At the working level of an IPM program, probably the single most important reason for attempting to get away from the use of broad-spectrum insecticides is to prevent the resurgence of the treated species and secondary pest outbreaks. There are two possible reasons for pest resurgences: (1) the reduction of natural enemies by the insecticide along with the pest and (2) removal of competitive species. Most explanations offered for pest resurgences or secondary pest outbreaks tend to support the first reason, destruction of beneficial insects.

The problems that result from blanket applications of broad-spectrum insecticides on agroecosystems are apparent everywhere. In 1966 scheduled applications of carbaryl were made in the Imperial valley of California in an effort to eradicate the pink bollworm. This resulted in destructive outbreaks of the cotton leafperforator and to a lesser extent the spider mite. Natural enemies of many pests were eliminated, and later, when the cotton treatments were stopped, severe outbreaks of cabbage loopers and beet armyworms developed on fall-planted crops. Although eradication was not attempted when the pink bollworm completed its spread across Arizona in 1965, scheduled applications of insecticide to control this pest since 1966 have also produced severe outbreaks of cotton leafperforator.

Value of selective versus broad-spectrum insecticides

The adverse consequences of using broad-spectrum insecticides without regard to all components of the environment have been well documented. Alternative methods have been sought to minimize these effects, including increased emphasis on selective insecticides, particularly the use of insecticides to gain selective action. The phenomenon of selectivity must be utilized not only from the standpoint of protecting beneficial species but also with regard to maximum effectiveness against the pest species. Methyl parathion, applied at the rate of 1.0 pound per acre, is quite effective against the cotton bollworm and the tobacco budworm in Arizona whereas ethyl parathion is relatively ineffective. Unfortunately, a popular formulation consisting of 6 pounds of ethyl and 3 pounds of methyl parathion per gallon is marketed in that area. Many applications of this formulation, at the rate of one pint or 1.125 pounds of total toxicant per acre, have been made to cotton for cotton bollworm or tobacco budworm control. Results have been disastrous because the 0.375 pound of methyl parathion was insufficient to provide control, and the total dosage of both eliminated beneficial species.

Achieving selective action with insecticides has not been easy. In most cases it has required a more detailed knowledge not only of the key pests but of the secondary pests and beneficial species as well. Selectivity has been achieved by a number of means, but in general it has been obtained by one of the following practices: (1) use of truly selective materials which are inherently more toxic to some species than to others, (2) timing of applications to suppress pest species while having minimum detrimental effects on natural enemies, and (3) reduced dosages to give adequate control of the target pest while sparing relatively large numbers of natural enemies of the target pest or other potential pests or both.

3.85 How has selectivity been achieved?

---

1. use of a chemical which is more toxic to certain insects
2. timing of applications
3. reduced dosages

In apple orchards in Nova Scotia, general use of a program of selective sprays for several years was followed by a reduction in the amount of damage caused by the key pests. DDT was replaced by ryania, and not only was good control of the codling moth obtained but large numbers of beneficial insects and mites in several major groups were preserved. Progress was also made by reducing the dosage of some of the more toxic chemicals to levels where pest species were controlled and beneficial species were conserved.

Significant advances have been made in IPM in Washington and Oregon on fruit-tree pests. Studies on apples have shown a wide selection of insecticides that would allow the predaceous mite *Typhlodromus occidentalis* Nesbitt to survive in moderate numbers. Even with some of the highly toxic materials, it was found that predator mortality could be reduced by using selective spray techniques or in some cases selective timing of sprays. Standard spray programs to control other pests, particularly the codling moth, ultimately resulted in severe spider mite problems. Acaricides that usually controlled the McDaniel mite became ineffective because of mite resistance to them. It was found that under the standard spray program, mite control measures were needed more frequently, at higher dosages, and later in the season. The primary function of the acaricide as used here delayed rather than controlled the development of mite populations. High population at harvest reduced fruit size, produced poor fruit color, and possibly affected fruit maturity. It was found that in the integrated control program, the McDaniel mite was practically nonexistent, or reached its peak, usually at lower levels, and declined well before harvest.

Similar problems were encountered in developing an integrated control program on pears in Oregon. Among the many problems which had to be considered were the low population levels of the pear rust mite and the two-spotted spider mite producing economic injury. These low economic injury levels tended to discourage pear growers from allowing the time necessary for the typical lag period between buildup of mites and their predators.

Another problem was the intense, concurrent summer spray programs for control of the pear psylla and the codling moth. One of these pest problems was solved when an area-wide dormant spray program for pear psylla control replaced the multiple summer sprays. The remaining problem was the commercial practice of codling moth control with multiple applications of azinphosmethyl at rates of 1.5 to 2.0 pounds per acre. Insecticidal control of the codling moth was essential to prevent economic losses; however, the dosages used eliminated predators. These points were demonstrated in an orchard that had received standard commercial treatment until 1964, at which time all summer sprays were omitted. Two-spotted spider-mite population densities peaked at 17, 2.5, 2.5, and 0.18 mites per leaf, respectively, for the succeeding four years. This showed that when pesticide applications were stopped, a trend toward lower average population densities of the two-spotted mite and corresponding increases in predatory mite levels followed. During this period, however, the codling moth caused fruit damage averaging about 40 percent, demonstrating the need for chemical control.

Subsequent studies with lower rates of azinphosmethyl showed that the reduced rate of 0.5 pound per acre gave adequate codling moth control but still eliminated predatory mites. Still lower rates of 0.125 and 0.25 pound per acre improved survival of the predatory mite, but gave inadequate codling moth control. In 1966, it was found that azinphosmethyl at 0.125 pound per acre in combination with oil provided the best overall control with good predator mite survival. Although codling moth control was not acceptable, it did warrant further testing in commercial orchards where moth levels would be considerably lower than encountered in the test orchard. Under those conditions the reduced rates of azinphosmethyl were adequate for control of the codling moth and were selective to the point of allowing increased densities of the predatory mite. During the first year of the study, phytophagous mites increased to unacceptable levels, but thereafter a more favorable predator-prey ratio developed which prevented such increases. This demonstrated the conversion phase of changing from the standard to the integrated program, in which the plant-feeding mites exceeded the economic level the first year. This again demonstrates the complexity of agroecosystems and the fact that changes from complete reliance upon chemicals cannot be made instantaneously.

3.86 What three factors were important in developing an integrated control program on pears?

- - - - - - - - - - - - - - - -

1. reevaluation of economic levels of plant-feeding mites
2. dormant spray to control pear psylla
3. lower dosage of insecticide to control codling moth

Methods of improving selectivity

So far we have learned that many insecticides do exhibit differential toxicities to various insects and mites, particularly among insect or mite groups. From a practical standpoint, however, little of this inherent selectivity is realized when insecticide dosage levels considered necessary to control major pests are applied on a field basis, often with repeated applications. This was indicated in research on tree-fruit insects where much of the selective action of the insecticides was achieved through changes in timing, numbers, and dosages of insecticide applications, accompanied by reevaluations of economic levels and influences of predators on their prey.

With the current difficulties in obtaining registrations for new pesticides it would take several years before new, truly selective insecticides could be available for grower use, assuming they were in existence today! It appears, therefore, that the greatest opportunity of using selective insecticides, at least in the near future, lies with agricultural entomologists who can devise ways of using the insecticides currently available in a manner more compatible with the total environment, including agricultural and all other human-centered ecosystems, than has been done in the past.

A number of investigators have demonstrated the possibilities of gaining selective advantage in insecticide use. Other approaches to more effective uses of insecticides are beginning to appear in the literature. Efforts are being made to apply greater proportions of the insecticides needed directly on the target crop and less on the surrounding areas. Significant advances in this area alone could result in reduced dosages and still control target pests. In Arizona it was shown that at all distances downwind, aerial applications resulted in four to five times as much insecticide drift as did applications from high-clearance ground sprayers. It was further shown that aerially applied insecticides apparently deposited less than 50 percent of the materials on target during the normal insecticide-use growing season. The addition of certain spray thickeners to insecticide emulsions, however, reduced drift from aerial applications. The benefit of these results from the standpoint of improving selectivity within the target area might be minimal. However, from the standpoint of protecting beneficial species or reducing residues in adjacent nontarget areas, the benefits might be manifold.

More precision in the placement and timing of insecticide applications offers hope for gaining selective action. Currently, the pink bollworm in Arizona necessitates repeated spray applications to prevent economic losses in the cotton crop. A decade ago, when it became a general problem across the state, control recommendations were indefinite as to when applications were needed; dosage levels of insecticides were excessive, for example, azinphosmethyl was used at 1.0 pound per acre. In recent years, with more precision in timing of applications, based on better estimates of economic population and damage levels, and research showing that 0.5 rather than 1.0 pound per acre of azinphosmethyl gives adequate control, maximum protection is now possible with a considerable reduction in the amount of insecticide required.

An experiment was conducted using insecticides for pink bollworm control in a much more selective way. This study involved the behavioral characteristics of both the pink bollworm and a sporadic pest complex including the cotton bollworm and the tobacco budworm. The pink bollworm oviposits on bolls in preference to squares, while the bollworm complex prefers the tender terminal growth. Newly-hatched cotton bollworm larvae feed for several days on squares in the terminal area before moving down the plant to attack bolls. Studies on population growth of the pink bollworm indicate that, in general, economic infestation levels are reached during the second boll generation, which is usually the latter part of July. This corresponds quite closely to the time when cotton bollworm and tobacco budworm oviposition begins. Thus, scheduled applications of insecticides are needed to control the pink bollworm at the time when predatory action against potentially damaging cotton bollworm populations is most needed. In many cases, cotton bollworm outbreaks occur following the initiation of pink bollworm treatments.

The experiment involving selective placement of the broad-spectrum insecticide azinphosmethyl demonstrated the possibility of controlling pink bollworms while sparing predators in the terminal portion of the cotton plant. This involved blocking out the over-the-row spray nozzle and utilizing only the two side nozzles to restrict the spray coverage to the lower two thirds of the plant where most of the bolls are located when insecticide applications are needed to control the pink bollworm. The minute pirate bug, an efficient cotton bollworm egg predator, was particularly abundant in the terminal area of the plants, and these, as well as other predatory species, were maintained in the plant terminals at levels comparable to the untreated checks. Pink bollworm control, though adequate, was not as good as in the three-nozzle/row treatment.

The potential for obtaining selective action with currently available insecticides appears to be great. Only a few of the ways have been mentioned. Other possibilities include insecticide-baited traps in conjunction with an attractant, chemically induced sterility, or even the use of other generally more specific types of insecticides, for example, insect pathogens.

3.87 How may selectivity be improved?

- - - - - - - - - - - - - - - -

1. insecticide-baited traps in conjunction with attractants
2. chemically induced sterility
3. more specific insecticides, for example, insect pathogens

Regardless of the means of achieving selectivity, little progress can be made in obtaining selective insecticidal action unless we have an adequate field force of well-trained entomologists schooled in concepts of insect pest management rather than "pest control." These persons would be exceptional individuals with intimate knowledge of crops, pest and beneficial species complexes, economic levels, relative toxicities of available insecticides to pest and beneficial species, appropriate formulations, application methods, and minimum dosages. It may be argued that we have this type of person already on the job and that his or her efforts allegedly indicate that as yet we are not achieving the goal of selectivity. This is not the case, at least in the numbers of qualified persons required to cover adequately the broad areas and many crops involved in the seasonal interactions between pests and their natural enemies. IPM specialists are now being trained in various departments of entomology across the country who should soon be available to help with this problem. There must be a more appreciative demand for this kind of person if the supply is to continue. This means selling the grower on the IPM approach to crop protection. It also means that this scientifically and economically sound approach toward meeting and reducing the problems and tensions of the present pesticide crisis must be more widely supported by all levels of the chemical industry, from top management to retail sales forces.

3.88 What should be included in the training of professional pest management specialists?

- - - - - - - - - - - - - - - -

1. crops in the agroecosystem
2. biology and ecology of both pest and beneficial insects
3. insecticides

References

Bartlett, B. R. 1964. Integration of chemical and biological control. Ch. 17 in Biological control of insect pests and weeds, ed. P. DeBach. Reinhold Publishing Corporation, New York.

Chant, D. A. 1964. Strategy and tactics of insect control. Can. Entomol. 96:182-201.

Gasser, R. 1966. Use of pesticides in selective manners. Proceedings of the FAO Symposium on Integrated Pest Control 2:109-113. Rome, 11-15 Oct. 1965.

Getzin, L. W. 1960. Selective insecticides for vegetable leaf-miner control and parasite survival. J. Econ. Entomol. 53(5):872-75.

Hoyt, S. C. 1969. Integrated chemical control of insects and biological control of mites on apples in Washington. J. Econ. Entomol. 62(1):74-86.

Madsen, H. F. and M. M. Barnes. 1959. Pests of pear in California. Calif. Agr. Exp. Sta. Ext. Serv. Circ. 478. 40 pp.

Metcalf, R. L. 1966. Requirements for insecticides of the future. Proceedings of the FAO Symposium on Integrated Pest Control 2:115-133. Rome, 11-15 Oct. 1965.

Newsom, L. D. 1966. Essential role of chemicals in crop protection. Proceedings of the FAO Symposium on Integrated Pest Control 2:95-108. Rome, 11-15 Oct. 1965.

Pickett, A. D. 1959. Utilization of native parasites and predators. J. Econ. Entomol. 52(6):1103-1105.

Ripper, W. E. 1956. Effect of pesticides on balance of arthropod populations. Ann. Rev. Ent. 1:403-438.

Slosser, J. E. and T. F. Watson. 1972. Population growth of the pink bollworm. Ariz. Exp. Sta. Tech. Bull. 195, 32 pp.

Smith, R. F. 1970. Pesticides: Their use and limitations in pest management. In Concepts of Pest Management, ed. R. L. Rabb and F. E. Guthrie. North Carolina State University, Raleigh, N.C.

Stern, V. M. 1966. Significance of the economic threshold in integrated pest control. Proceedings of the FAO Symposium on Integrated Pest Control 2:41-56. Rome, Oct. 1965.

Stern, V. M., R. F. Smith, R. van den Bosch and K. S. Hagen. 1959. The integrated control concept. Hilgardia 29(2):81-101.

Van den Bosch, R. and V. M. Stern. 1962. The integration of chemical and biological control of arthropod pests. Ann. Rev. Entomol. 7:367-386.

Ware, G. W., W. P. Cahill, P. D. Gerhardt, and J. M. Witt. 1970. Pesticide drift IV. On-target deposits from aerial application of insecticides. J. Econ. Entomol. 63(6):1982-1983.

Ware, G. W., B. J. Estesen, W. P. Cahill, P. D. Gerhardt, and K. R. Frost. 1969. Pesticide drift I. High-clearance vs. aerial application of sprays. J. Econ. Entomol. 62(4):840-843.

Ware, G. W., B. J. Estesen, W. P. Cahill, P. D. Gerhardt, and K. R. Frost. 1970. Pesticide drift III. Drift reduction with spray thickeners. J. Econ. Entomol. 63(4):1314-1316.

Watson, T. F. and D. G. Fullerton. 1969. Timing of insecticidal applications for control of the pink bollworm. J. Econ. Entomol. 62(3):682-685.
Westigard, P. H. 1969. Timing and evaluation of pesticides for control of the pear rust mite. J. Econ. Entomol. 62:1158-1161.
Westigard, P. H. 1971. Integrated control of spider mites on pear. J. Econ. Entomol. 64(2):496-501.

Legal Aspects of Insecticide Use

There are several federal laws protecting the users of insecticides, their pets and domestic animals, their neighbors, and the consumers of treated products. Today nothing is left unprotected.

In the beginning, the Federal Food, Drug, and Cosmetic Act of 1906, known as the Pure Food Law, was enacted requiring that food (fresh, canned, and frozen) shipped in interstate commerce be pure and wholesome. Nothing in the law pertained to insecticides. In 1910 the Federal Insecticide Act, the first legislation to control insecticides, was signed. The act covered only insecticides and fungicides and was designed mainly to protect the farmer from substandard or fraudulent products, for there were many.

3.89 The first federal law to control insecticides was the____________________.

- - - - - - - - - - - - - - -

Federal Insecticide Act of 1910

The Pure Food Law of 1906 was amended in 1938 to include pesticides on foods, primarily the arsenicals such as lead arsenate and Paris green. It also required the adding of color to white insecticides, including sodium fluoride and lead arsenate, to prevent their use as white look-alike cooking materials. This was the first federal effort to protect the consumer from pesticide-contaminated food by providing tolerances for pesticide residues, namely, arsenic and lead in foods where these materials were necessary for the production of a food supply.

3.90 Insecticide residues were first controlled on foods in 19__ by the amended __________________ Law.

- - - - - - - - - - - - - - -

38 Pure Food

The Federal Insecticide, Fungicide, and Rodenticide Act (FIFRA) became law in 1947; it superseded the 1910 Federal Insecticide Act and extended the coverage to include herbicides and rodenticides. It required that all these products be registered with the U.S. Department of Agriculture before marketing in interstate commerce. Basically, the law was one requiring good and useful labeling, making the product safe to use if label instructions were followed. The label was required to contain the manufacturer's name and address, the name, brand, and trademark of the product, its net contents, an ingredient statement, an appropriate warning statement to prevent injury to humans, animals, plants, and useful invertebrates, and directions for use adequate to protect the user and the public.

3.91 The first federal pesticide labeling law was the ______________________.

- - - - - - - - - - - - - - - - -

FIFRA of 1947

The Food, Drug, and Cosmetic Act (1906, 1938) was modified in 1954 by the passing of the Miller Amendment. It provided that any raw agricultural commodity may be condemned as adulterated if it contains any pesticide chemical whose safety has not been formally cleared or that is present above tolerances. In essence, this clearly set tolerances for all insecticides in food products, for example, 7.0 ppm DDT in lettuce, or 1.0 ppm ethyl parathion on string beans.

3.92 The 1954 ____ established the legal limits of pesticide residues (tolerances) which may appear in and on foods.

- - - - - - - - - - - - - - - - -

Miller Amendment of the Food, Drug, and Cosmetic Act

In practical operation these two federal statutes—the Federal Insecticide, Fungicide, and Rodenticide Act (FIFRA) and the Miller Amendment to the Food, Drug, and Cosmetic Act—supplement each other and are interrelated by law. Today they serve as the basic elements of protection for the applicator, the consumer of treated products, and the environment, as modified by the following amendments.

The Food Additives Amendment to the Food, Drug, and Cosmetic Act (1906, 1938, 1954) was passed in 1958. It extended the same philosophy to all food additives that had been applied to pesticide residues on raw agricultural commodities by the 1954 amendment. However, this also controlled pesticide residues in processed foods that had not previously fitted into the 1954 designation of raw agricultural commodities. Of greater importance, however, was the inclusion of the Delaney Clause, which states that any chemical found to be a carcinogen (a substance causing cancer) in laboratory animals when fed at any dosage may not appear in foods consumed by humans. This has become the most controversial segment of the federal laws applying to pesticides, since dosage limits are not set for such tests with experimental animals.

3.93 Known carcinogens cannot appear as residues in foods consumed by humans as set forth in 19____ by the ______________________________.

- - - - - - - - - - - - - - - -

58 Delaney Clause of the Food Additives Amendment

In 1959, the FIFRA (1947) was amended to include all other pesticides not previously included. Because the FIFRA and the Food, Drug, and Cosmetic Act are allied, these additional economic poisons are also controlled as they pertain to residues in raw agricultural commodities. So far, the various statutes mentioned apply only to commodities shipped in interstate commerce.

3.94 Other pesticides included under the FIFRA in 1959 were ______________, ______________, ______________, and ______________.

- - - - - - - - - - - - - - - -

nematicides plant regulators defoliants desiccants

The FIFRA (1947, 1959) was again amended in 1964 to require that all pesticide labels contain the Federal Registration Number. It also required caution words such as WARNING, DANGER, CAUTION, and KEEP OUT OF REACH OF CHILDREN to be included on the front label of all poisonous pesticides. Manufacturers were also required to remove safety claims from all labels.

The administration of the FIFRA had been the responsibility of the Pesticides Regulation Division of the U.S. Department of Agriculture until 1970. At that time the responsibility was transferred to the newly designated EPA. Simultaneously, the authority to establish pesticide tolerances was transferred from the Food and Drug Administration (FDA) to the EPA. The enforcement of tolerances remained the responsibility of the FDA.

The FIFRA (1947, 1959, 1964) was revised in 1972 by the most important piece of pesticide legislation of this century. The Federal Environmental Pesticide Control Act (FEPCA), sometimes referred to as the 1972 FIFRA Amendment. The significant provisions of the FEPCA are condensed as follows:

1. Use of any pesticide inconsistent with the label is prohibited.
2. Deliberate violations of the FEPCA by growers, applicators, or dealers can result in heavy fines or imprisonment or both.
3. All pesticides will be classified into (a) general-use or (b) restricted-use categories by October 1976.
4. Anyone applying restricted-use pesticides must be certified by the state in which he or she lives.
5. Pesticide manufacturing plants must be registered and inspected by the EPA.
6. States may register pesticides on a limited basis when intended for special local needs.
7. All pesticide products must be registered by the EPA, whether shipped in interstate or intrastate commerce.
8. For a product to be registered, the manufacturer is required to provide scientific evidence that the product, when used as directed, will (1) effectively control the pests listed on the label, (2) not injure humans, crops, livestock, and wildlife, or damage the total environment, and (3) not result in illegal residues in food or feed.

Ten categories of certification for commercial applicators were established: (1) agricultural pest control (plant and animal), (2) forest pest control, (3) ornamental and turf pest control, (4) seed treatment, (5) aquatic pest control (6) right-of-way pest control (7) industrial, institutional, structural, and health-related pest control, (8) public health pest control, (9) regulatory pest control, and (10) demonstration and research pest control.

The FEPCA also set general standards of knowledge for all categories of certified commercial applicators. Testing will be based, among other things, on the following areas of competency: (1) label and labeling comprehension, (2) safety, (3) environment, (4) pests, (5) pesticides, (6) equipment, (7) application techniques, and (8) laws and regulations.

These are only the most important aspects of the FEPCA that you, the interested novice, need be acquainted with.

3.95 Describe the eight basic provisions of the FEPCA as briefly as possible.

\- - - - - - - - - - - - - - -

- - - - - - - - - - - - - - - -

(1) The label must be followed to a "T," (2) violators are punished, (3) pesticides are classed as general or restricted use, (4) restricted use pesticides are to be applied only by a certified applicator, (5) pesticide manufacturing plants must be registered, (6) states may register pesticides for local needs, (7) all pesticides must be registered by EPA, and (8) certain data must be provided for registration of a pesticide.

Each state usually has two or three similar laws controlling the application of pesticides and the sale and use of pesticides. They may or may not involve the licensing of aerial and ground applicators.

3.96 The most significant pesticide law of this century is the ______________________________.

- - - - - - - - - - - - - - - -

1972 Federal Environmental Pesticide Control Act (FEPCA)

3.97 Pesticides are classified into two categories by the FEPCA, ______________ and ________________.

- - - - - - - - - - - - - - - -

general use   restricted use

3.98 Anyone applying a restricted-use pesticide must be ____________________________.

- - - - - - - - - - - - - - - -

certified

The Insecticide Label

<u>The FEPCA and the Label</u>. The FEPCA contains three very important points concerning the pesticide label which we feel should be brought further to your attention and emphasized. They pertain to reading the label, understanding the label directions, and following these instructions carefully.

Two of the first provisions of the FEPCA are that the use of any pesticide inconsistent with the label is prohibited, and deliberate violations by growers, applicators, or dealers can result in heavy fines or imprisonment or both. The third provision is found in the general standards for certification of commercial applicators to use restricted-use pesticides, the area of label and labeling comprehension. Applicators will be tested on (a) the general format and terminology of pesticide labels and labeling; (b) the understanding of instructions, warnings, terms, symbols, and other information commonly appearing on pesticide labels; (c) classification of the product (general or restricted use); and (d) the necessity for use consistent with the label.

In Figures 3.16 and 3.17 are shown the format labels for general-use and restricted-use pesticides as currently proposed by the EPA Registration Division. At this writing these formats are only proposed, but will very likely become required beginning in late 1975. These labels are keyed as follows:

- 1. Product name
- 2. Company name and address
- 3. Net contents
- 4. EPA pesticide registration number
- 5. EPA formulator or manufacturer establishment number
- 6A. Ingredients statement
- 6B. Pounds/gallon statement (if liquid)
- 7. Front panel precautionary statements
- 7A. Child hazard warning, "Keep Out of Reach of Children"
- 7B. Signal word — DANGER, WARNING, or CAUTION
- 7C. Skull and crossbones and word "Poison" in red
- 7D. Statement of practical treatment
- 7E. Referral statement
- 8. Side/back panel precautionary statements
- 8A. Hazards to humans and domestic animals
- 8B. Environmental hazards
- 8C. Physical or chemical hazards
- 9A. Restricted Use Pesticide block
- 9B. Statement of pesticide classification
- 9C. Misuse statement
- 10A. Re-entry statement
- 10B. Category of applicator
- 10C Storage and Disposal block
- 10D Directions for use

Figure 3.16 EPA format for General Use pesticide label

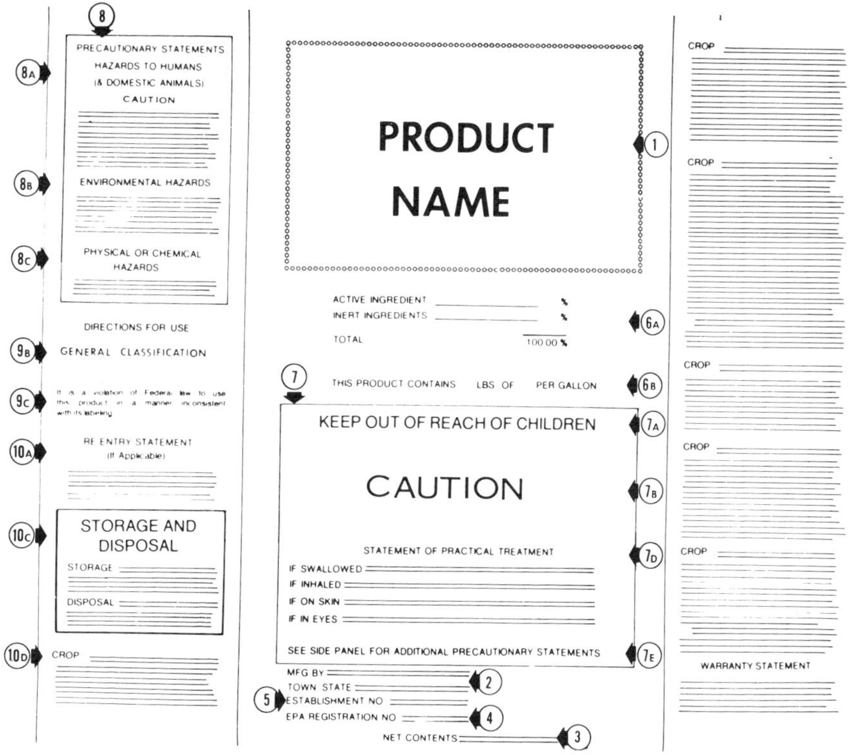

Figure 3.17 EPA format for Restricted Use pesticide label

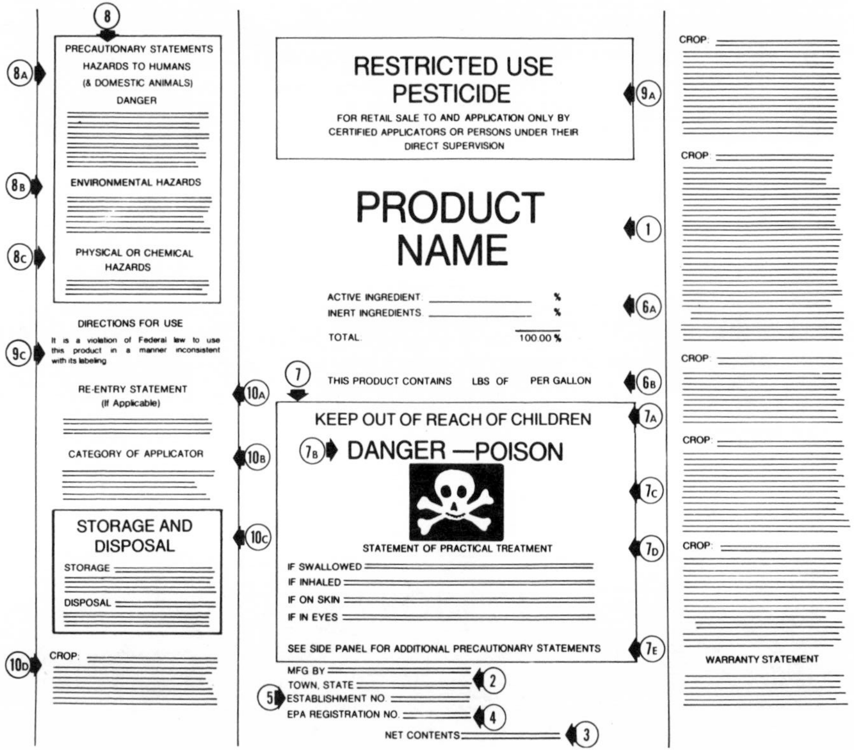

3.99 Who can apply Restricted Use pesticides?

- - - - - - - - - - - - - - - -

Only Certified Applicators. (9A)

3.100 Who can apply General Use pesticides?

- - - - - - - - - - - - - - - -

Anyone.

3.101 What segment of either label is personally most important to the applicator?

- - - - - - - - - - - - - - - -

Precautionary and practical treatment statements as they pertain to his own safety and health. (7D, 8A)

3.102 Where would you look to determine if the product could be used on beef cattle or field tomatoes?

- - - - - - - - - - - - - - - -

In the Directions for Use. (10D)

3.103 What is meant by 10A, Re-entry Statement?

- - - - - - - - - - - - - - - -

Waiting period after crop is treated before workers are permitted to enter. (See page 126).

3.104 Can a pesticide be used on a crop not shown on the label?

- - - - - - - - - - - - - - - -

Absolutely not! (9C)

Field Reentry Safety Intervals

The EPA now requires safety waiting intervals between application of certain insecticides and worker reentry into all treated fields to prevent unnecessary exposure. Several states (for example, California) have adopted waiting intervals longer than those required by the EPA. The waiting intervals established by EPA are

Ethyl parathion	48 hours
Methyl parathion	48 hours
Demeton (Systox).	48 hours
Monocrotophos (Azodrin).	48 hours
Carbofenothion (Trithion).	48 hours
Oxydemetonmethyl (MetaSystox-R) . .	48 hours
Dicrotophos (Bidrin)	48 hours
Endrin	48 hours
Azinphosmethyl (Guthion)	24 hours
Phosalone (Zolone)	24 hours
EPN	24 hours
Ethion	24 hours

For all other insecticides it is necessary only that workers wait until sprays have dried or dusts have settled before reentering treated fields. These worker safety intervals are not to be confused with the familiar harvest intervals, the minimum days from last treatment to harvest, indicated on the insecticide label.

If it is necessary for workers to enter fields earlier than the required waiting intervals, they must wear protective clothing. This consists of a long-sleeved shirt, long-legged trousers or coveralls, hat, shoes, and socks.

These waiting intervals should not impose any undue hardship on pest-management specialists, because application of any one of the above materials would preclude the necessity for field inspection within the required waiting intervals.

***

3.105 If the insecticide in question is not included in the list above, the reentry waiting interval will automatically be to ______________________________ ______________________________.

\- - - - - - - - - - - - - - - -

wait until sprays have dried or dusts have settled

***

3.106 When a field treated with one of the insecticides listed above must be entered by a pest-management specialist prior to the established waiting interval, what precautions must be taken?

\- - - - - - - - - - - - - - - -

He must wear protective clothing as described.

Pesticide Emergencies

All insecticides can be used safely, provided common-sense safety is practiced and provided they are used according to the label instructions; this includes keeping them away from children and illiterate or mentally incompetent persons. Despite the most thorough precautions, accidents will occur. Below are given two important sources of information in the event of any kind of serious pesticide accident.

The first and most important source of information is the CHEMTREC telephone number. From this toll-free long-distance number can be obtained emergency information on all pesticide accidents, pesticide-poisoning cases, pesticide spills, and pesticide-spill cleanup teams. This telephone service is available twenty-four hours a day. The toll-free number is

CHEMTREC 800-424-9300

CHEMTREC is the abbreviation for Chemical Transportation Emergency Center and is sponsored by the Manufacturing Chemists Association (MCA) in cooperation with the Pesticide Safety Team Network (PSTN).

3.107 Any kind of pesticide emergency information can be obtained from the national toll-free telephone service known as ________________.

CHEMTREC

The second source of information is only for human-poisoning cases: It is the nearest Poison Control Center. Look it up in the telephone directory under POISON CONTROL CENTERS, or ask the telephone operator for assistance. Poison Control centers are usually located in the larger hospitals of most cities and can provide emergency treatment information on all types of human poisoning, including pesticides. The telephone number of the nearest Poison Control Center should be kept as a ready reference by every IPM specialist for the protection of all who work with insecticides.

Specific pesticide-poisoning information can be obtained in writing or by telephone from

National Clearing House for
Poison Control Centers
HEW, Food & Drug Administration
Bureau of Drugs
5401 Westbard Avenue
Bethesda, Maryland 20016

3.108 Where in the nearest city can you obtain pesticide-poisoning information?

- - - - - - - - - - - - - - - -

From Poison Control centers.

Safe Handling of Insecticides

"Safety is a state of mind" is an old cliché in the rule of industrial safety engineers. Pesticide safety is more than a state of mind. It must become a habit with those who handle, load, apply, mix, haul, and sell pesticides—and especially with those who supervise those who do!

There surely must be literally thousands of "do" and "don't" rules concerning the safe handling of pesticides. We have gleaned these and surfaced with the ten we believe most important.

1. Inform all employees of the hazards associated with pesticide use and have prearranged medical services available.
2. Post telephone numbers of attending physician and the nearest Poison Control Center at every telephone in the work situation.
3. Apply insecticides only when needed for a specific purpose and only as recommended. FOLLOW THE LABEL.
4. Avoid absorption of insecticides through the skin, mouth, or lungs.
5. Wear protective clothing (laundered after each use), rubber gloves, and a suitable respirator when mixing, pouring, or applying insecticides belonging to the "Highly Toxic" classification.
6. Destroy insecticide containers as soon as they are emptied. Empty containers should be made unusable by breaking or crushing and then disposing in an approved manner.
7. Store unused insecticides in original containers beyond the reach of children and animals. Lock them up.

8. Never use spray equipment which has been used for 2,4-D or other herbicides.
9. Clean up spills immediately. Wash pesticides off skin promptly with plenty of soap and water. Change clothes immediately if they become contaminated.
10. Avoid damage to beneficial and pollinating insects by not spraying during periods when such insects are actively working in the spray area. Notify neighboring beekeepers, as required by legislative regulations, at least twenty-four hours before application so that precautionary measures can be taken.

3.109 In the list of ten rules pertaining to the safe handling of insecticides, which one is the most important to you in your work situation?

- - - - - - - - - - - - - - - -

Only you know!

3-D HOST-PLANT RESISTANCE

The resistance of plants to insect attack is the result of their having inherited qualities that determine the extent of insect damage. It should be recognized that host-plant resistance is not a cure-all, but is an important IPM component which must be fitted carefully to the control of specific insects and into crop-improvement programs.

3.110 Host-plant resistance is the result of ________________ qualities of plants that determine the extent of insect __________.

- - - - - - - - - - - - - - - -

inherited damage

Classification of Resistance

The degree of pest resistance ranges from varieties with *immunity*, where a specific insect never consumes or injures the plants under any known conditions, to varieties with *high susceptibility* that suffer much more than average damage by a particular insect. Terms used to classify resistance between the two extremes are high resistance, low resistance, and susceptibility.

3.111 The extremes in degree of plant resistance to insect attack are ______________________ and ______________________.

immunity high susceptibility

Mechanisms of Resistance

Knowledge of the mechanisms, or basis, of resistance of plants to insects is very important in understanding the place and capacity of resistance in the IPM system. The three mechanisms of resistance are *antibiosis*, *tolerance* and *preference* or *nonpreference*. These mechanisms are often illustrated in the form of a triangle as shown below. Any or all of these mechanisms may be responsible for resistance of a variety to a particular insect.

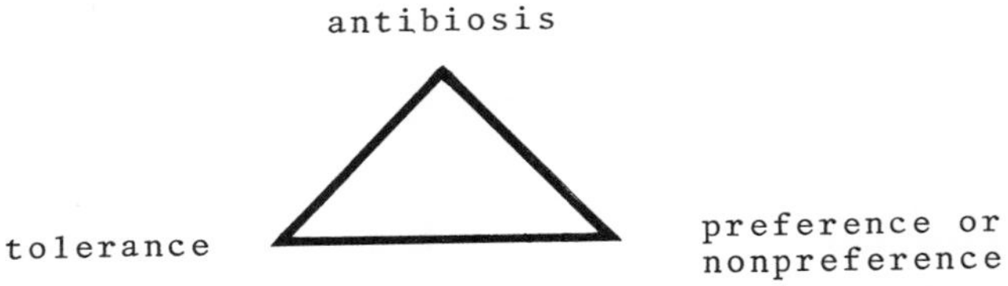

Antibiosis is the capacity of a variety to prevent, injure, or destroy insect life. It accounts for the adverse effects on an insect's life history when it feeds on a resistant variety. The effects on the insect are reduced fecundity, decreased size, abnormal length of life, and increased mortality.

Tolerance is a basis of resistance in which the plant shows an ability to grow and reproduce or repair injury in spite of supporting a population of insects that would damage a susceptible variety.

Preference or nonpreference refers to the group of plant characters and insect responses that lead to or away from the use of a particular plant or variety for oviposition, food, or shelter, or for a combination of the three.

3.112 The three mechanisms of resistance are __________, __________, and ______________________.

- - - - - - - - - - - - - - - - -

antibiosis tolerance preference or nonpreference

Apparent Resistance

Apparent resistance is the result of temporary characters in potentially susceptible host plants. Varieties which demonstrate this type of resistance are of considerable importance in IPM, but they should be distinguished from varieties which are resistant under a wider range of conditions. Three types of apparent resistance are *host evasion, induced resistance,* and *escape.*

A variety may evade damage by passing through the susceptible stage quickly or at a time when insect numbers are low. Some hosts may mature early, for example, before a potentially damaging insect reaches an economic level; thus the hosts evade high populations which would do serious damage to them had they matured late. Several early-maturing cotton varieties evade late-season pink bollworm populations in this manner.

Induced resistance refers to temporarily increased resistance resulting from some condition in the plant or environment such as a change in the amount of water or soil fertility. Plant-bug populations are often lower in fields of cotton and other crops with poor moisture than in those with good moisture.

Escape refers to the lack of infestation of, or injury to, the host plant because of transitory circumstances such as incomplete insect infestation. Even though not resistant, some plants may remain uninfested in fields with high insect populations because they escape infestation for some unexplained reasons.

3.113 Three types of apparent resistance are ________, ________ and ________.

- - - - - - - - - - - - - - - - -

host evasion induced resistance escape

Relationship of Host-Plant Resistance to IPM

Host-plant resistance, as a component of the IPM system, can serve as the principal control method or as an adjunct to other control measures. Work in host-plant resistance may also serve as a safeguard against the release of susceptible varieties. The susceptible varieties may be identified prior to release in resistance studies.

3.114 Host-plant resistance, as a component of the IPM system, can serve as the ________ of control, as an ________________ to the other control measures, or as a ____________ against release of susceptible varieties.

- - - - - - - - - - - - - - - - -

principal method adjunct safeguard

An outstanding example of resistance as the principal control method is the use of resistant rootstock of American grapes as a means of controlling grape phylloxera on European grapes. Although discovered more than a century ago this method of controlling grape phylloxera still exists and points to the permanence involved in host-plant resistance. Other examples, such as Hessian fly resistance in wheat, also show the value and potential of host-plant resistance as the principal method of control. This successful and permanent type of resistance has been largely possible only in situations of high host specificity by the insect involved.

It is as an adjunct to other control measures that host-plant resistance will probably be of most use in IPM. A low level of resistance may be all that is needed to prevent economic loss when added to control by other methods. Chinch bugs have been successfully controlled in corn and sorghum by the use of resistant varieties and barriers designed to protect the more susceptible seedling plants. Spotted alfalfa aphid control in alfalfa has been attained by the use of resistant varieties and beneficial insects. The presence of low levels of aphids enables parasite and predator populations to maintain themselves. This is helpful not only in terms of aphid control but also from the standpoint of maintaining reservoirs of beneficial insects useful in preventing outbreaks of the pests in nearby fields of other crops.

3.115 In what way will host-plant resistance probably be of most use in IPM?

- - - - - - - - - - - - - - - - -

As adjunct to other control measures.

Limitations and Advantages of Host-Plant Resistance

Limitations to the use of host-plant resistance include the time required for developing resistant varieties; the development or selection of insect biotypes (defined below) capable of attacking resistant varieties; incompatibility of resistance characters with other desirable agronomic characters; and the general lack of acceptance and continued use of resistant varieties by growers. None of the limitations are serious enough to justify the lack of emphasis on developing and using resistant varieties as has occurred in the past. In some cases many years have been required to develop resistant varieties. As a parallel it is also true that many years are required to develop and put to use a new insecticide. Sometimes resistance can be utilized sooner when it is found in existing varieties. For example, alfalfa plants resistant to the spotted alfalfa aphid were found among existing varieties soon after the insect was discovered in the United States. These plants were quickly developed into varieties that have continued to provide excellent control of the insect despite the development of at least six aphid *biotypes* (insect adaptations which permit the population to do well on previously resistant varieties). Scientists have been able to alleviate the biotype problem by developing and releasing varieties resistant to the new biotype of a given area. It is unlikely that alfalfa could be raised commercially in the Southwest without varieties resistant to the spotted alfalfa aphid.

3.116 The development or selection of insect __________________ is one of the potential limitations to the use of host-plant resistance.

- - - - - - - - - - - - - - - -

biotypes

Incompatibility of factors for insect resistance with other desirable agricultural or economic characteristics is another possible limitation of resistant varieties. Only rarely has this been a problem, however, and it should not restrict plant-resistance studies. It is necessary to evaluate carefully the performance of a variety resistant to one pest to assure that other pests are not attracted or benefited in some way by the character causing resistance to the first pest.

The advantages of host-plant resistance as a control method in the IPM system are many. For example, insect control by resistance is *specific*, *cumulative*, and *persistent*. A variety that is only resistant enough to reduce an insect infestation by a small percentage may bring the pest to subeconomic levels within a few generations. Since many insects complete several generations each year, the beneficial effect may be dramatic during a single season. The cumulative beneficial effect is in direct contrast to the usual decline in effectiveness common to insecticides.

3.117 Control by host-plant resistance is an outstanding control method because it is __________________, __________________ and __________________.

- - - - - - - - - - - - - - - -

- - - - - - - - - - - - - - - -

specific cumulative persistent

Other important advantages of host-plant resistance include the low initial investment, usually with no additional costs to the grower, a complete lack of hazard to man or the environment, and compatibility with other IPM methods, especially the full utilization of parasites and predators.

References

Anonymous. 1969. Insect-pest management and control. Nat. Acad. Sci. Publ. 1695.

Maxwell, F. G. 1972. Use of plant resistance in pest control. Implementing practical pest management strategies: Proceedings of a National Extension Insect Pest Management Workshop. Purdue University, West Lafayette, Ind. Mar. 14-16.

Painter, R. H. 1951. Insect resistance in crop plants. The Macmillan Company, New York. 521 pp.

Painter, R. H. 1958. Host plant resistance to insects. Ann. Rev. Entomol. 3:267-290.

Painter, R. H. 1968. Crops that resist insects provide a way to increase world food supply. Kansas State Agr. Exp. Sta. Bull. 520.

3-E PHYSICAL AND MECHANICAL CONTROLS

Physical and mechanical controls are direct or indirect measures that kill the insect, disrupt its physiology by means other than insecticides, or adversely alter the insect's environment. They are different from cultural control in that the devices or action are directed against the insect instead of being a modification of some agricultural practice. For instance, the use of a fly swatter against annoying flies is a form of mechanical control, while hand-picking tomato hornworms from tomato plants is a form of physical control.

3.118 Physical and mechanical controls differ from other forms of control in that they ______________ the insect, ______________ its physiology, or change the insect's ______________ to make it unlivable.

- - - - - - - - - - - - - - - -

kill disrupt environment

These forms of control are the oldest and in some cases the most primitive of all. This fact may be deceptive, however, because many physical or mechanical forms of control are highly useful and effective, despite the more appealing and technologically new methods.

As much as any other component of IPM, physical and mechanical controls require a thorough knowledge of the pest's ecology and its biological weaknesses. Unfortunately, too few insect pests are well understood ecologically, and this has contributed to the fact that physical and mechanical controls have generally played only minor roles in IPM.

3.119 Why have physical and mechanical methods not been used more commonly in insect control?

- - - - - - - - - - - - - - - -

Not enough was known about the insect's relationship to its environment, that is, its ecology.

Forms of control within this category are utilizing high and low temperatures; reducing humidity; utilizing insect attraction to light traps; attracting, repelling or killing by sound; constructing barriers and excluders; picking by hand; shaking and jarring; herding; and trapping. Only light traps and barriers have been used with any degree of success in IPM, though the others may have become involved in a very minor way.

Excluders and barriers used in the past include dust ridges or trenches to prevent chinch bug migration into fields. When the bugs began their migration they were retained within the dust barriers. A similar principle has been applied to prevent salt marsh caterpillars from moving into uninfested fields. Instead of using a dust barrier, however, a 6-inch aluminum strip is placed vertically around field borders, imposing an unsurmountable obstacle to the caterpillars. Within hours, they can be found stacked against the strips, dead.

Diapausing larvae of the pink bollworm can be carried to uninfested cotton-growing areas in cotton seed. In a search for ways to reduce this route of movement it was observed that mechanical delinting of the seed reduces pink bollworm survival measurably, and that acid delinting is 100 percent effective. Though usually not classed as a physical or a mechanical control method, the sulfuric-acid delinting treatment has been a means of controlling seed-borne pink bollworms for several decades. Consequently, several cotton-producing states require acid delinting of cotton seed entering or leaving the state as a precaution against this costly pest.

The use of abrading dusts as stored-grain protectants could fall into the category of physical control. These dusts, such as diatomaceous earth, remove the surface waxes from the cuticle of stored-grain beetles and abrade their cuticle in a way that causes them to die from desiccation. Abrading dusts may also be referred to as sorptive dusts.

All of this brings us to the second most sensational and potentially useful of all control methods, ultraviolet (UV) light. (The most sensational, of course, is the phenomenal attractiveness of female sex pheromones to the males!)

3.120 Next to sex pheromones, what are many species of insects most attracted to?

\- - - - - - - - - - - - - - - - -

Selected wavelengths of light, namely, ultraviolet.

To varying degrees, thousands of insect species respond to light. Most of these photosensitive or phototactic species belong to the orders Ephemeroptera (mayflies), Neuroptera (nerve-winged insects), Orthoptera (grasshoppers, crickets, cockroaches), Hemiptera (true bugs), Coleoptera (beetles), Trichoptera (caddis flies), Lepidoptera (moths, butterflies), Diptera (flies), and Hymenoptera (bees, wasps).

Generally speaking, insects that are active during the day (diurnal) are not attracted significantly to artificial light. Most of the species that have a distinct phototactic response are most active at night, at dusk, or at dawn (nocturnal).

3.121 Generally speaking, what is the response of diurnal insects to ultraviolet light?

\- - - - - - - - - - - - - - - - -

They are not very responsive to any wavelength at night.

Several pest species in the order Lepidoptera are attracted to light. These include the codling moth, the Oriental fruit moth, the tobacco budworm, the cotton bollworm or corn earworm, various cutworms, the fall armyworm, the cabbage looper, the European corn borer, the pink bollworm, and the tobacco and tomato hornworms.

Many species respond only to certain wavelengths of light, and most show definite time peaks of nocturnal activity. Because of the enormity of the number of insects which are attracted to light, only a partial list of the common economic species has been assembled and appears in Table 3.5.

Table 3.5 Partial listing of the common economic insect species attracted to ultraviolet light.

COLEOPTERA	
Acalymma vittatum (F.)	striped cucumber beetle
Agonoderus lecontei Chd.	seed-corn beetle
Amiphimallon majalis (Razoumowsky)	European chafer
Bothynus gibbosus (DeG.)	carrot beetle
Conoderus falli Lane	southern potato wireworm
Conoderus vagus (Cand.)	wireworm
Conotrachelus nenuphar (Herbst)	plum curculio
Cyclocephala borealis Arrow	northern masked chafer
Diabrotica spp.	cucumber beetles
Diabrotica longicornis (Say)	northern corn rootworm
Elateridae	click beetles
Epicauta lemniscata (F.)	three-striped blister beetle
Epicauta vittata (F.)	striped blister beetle
Galerucella xanthomelaena (Schr.)	elm leaf beetle
Lasioderma serricorne (F.)	cigarette beetle
Maladera castanea (Arrow)	Asiatic garden beetle
Oryzaephilus mercator (Fauvel)	merchant grain beetle
Phyllophaga spp.	June beetles
Rhyzopertha dominica (F.)	lesser grain borer
Sitophilus oryzae (L.)	rice weevil
Tribolium castaneum (Herbst)	red flour beetle
DIPTERA	
Anopheles spp.	mosquitoes
Chaoborus astictopus Dyar & Shannon	Clear Lake gnat
Culex spp.	mosquitoes
Culicoides spp.	blood-sucking midges
Drosophila spp.	vinegar flies
Haematobia irritans (L.)	horn fly
Hippelates	eye gnats
Hylemya antiqua (Meig.)	onion maggot
Hylemya brassicae (Bouché)	cabbage maggot
Mansonia perturbans	mosquito
Musca domestica (L.)	house fly
Psychodidae	moth flies
Sepsidae	"black scavenger flies"
Simuliidae	biting gnats
Stomoxys calcitrans (L.)	stable fly
Tabanidae	deer and horse flies

HEMIPTERA	
Acrosternum hilare (Say)	green stink bug
Adelphocoris lineolatus (Goeze)	alfalfa plant bug
Adelphocoris rapidus (Say)	rapid plant bug
Labopidea allii Knight	onion plant bug
Lygus pallidulus (Blanch.)	lygus bug
Lygus lineolaris (P. deB.)	tarnished plant bug
Lygus pabulinus (L.)	lygus bug
Psallus seriatus (Reuter)	cotton fleahopper
Rhinacloa forticornis (Reuter)	black fleahopper

HOMOPTERA	
Aphidae	aphids
Cicadellidae	leafhoppers
Draeculacephala mollipes (Say)	leafhopper
Empoasca fabae (Harris)	potato leafhopper
Membracidae	treehoppers
Psylla pyricola Foerster	pear psylla

LEPIDOPTERA	
Acrobasis vaccinii Riley	cranberry fruitworm
Agrotis gladiaria (Morr.)	clay-backed cutworm
Agrotis malefida Guinée	pale-sided cutworm
Agrotis ipsilon (Hufnagel)	black cutworm
Alabama argillacea (Hübner)	cotton leafworm
Amathes c-nigrum (L.)	spotted cutworm
Anagrapha falcifera (Kirby)	celery looper
Archips argyrospilus (Walker)	fruit-tree leaf roller
Argyrotaenia valutinana (Walker)	red-banded leaf roller
Atteva aurea (Fitch)	ailanthus webworm
Autographa spp.	loopers
Bucculatrix canadensisella Chamb.	birch skeletonizer
Cadra cautella (Walker)	almond moth
Caenurgina crassiuscula (Haworth)	clover looper
Caenurgina erechtea (Cramer)	forage looper
Carpocapsa pomonella (L.)	codling moth
Celerio lineata (F.)	white-lined sphinx
Ceratomia catalpae (Bdvl.)	catalpa sphinx
Choristoneura fumiferana (Clam.)	spruce budworm
Crymodes devastator (Brace)	glassy cutworm
Diacrisia virginica (F.)	yellow woollybear
Elasmopalpus lignosellus (Zell.)	lesser cornstalk borer
Ephestia elutella (Hübner)	tobacco moth
Estigmene acrea (Drury)	salt-marsh caterpillar
Faronta diffusa (Walker)	wheat head armyworm
Feltia ducens (Walker)	cutworm

LEPIDOPTERA (Continued)	
Feltia subgothica (Haworth)	dingy cutworm
Feltia subterranea (F.)	granulate cutworm
Grapholitha packardi Zeller	cherry fruitworm
Harrisina americana (Guerin-Meneville)	grape leaf skeletonizer
Heliothis virescens (F.)	tobacco budworm
Heliothis zea (Boddie)	corn earworm, cotton bollworm
Isia isabella (J. E. Smith)	banded woollybear
Lacinipolia renigera (Stephens)	bristly cutworm
Loxostege similalis (Guenée)	garden webworm
Malacosoma americanum (F.)	eastern tent caterpillar
Manduca quinquemaculata (Haworth)	tomato hornworm
Manduca sexta (Johannson)	tobacco hornworm
Nephelodes emmedonius (Cramer)	bronzed cutworm
Ostrinia nubilalis (Hübner)	European corn borer
Papaipema nebris (Guenée)	stalk borer
Pectinophora gossypiella (Saunders)	pink bollworm
Peridroma saucia (Hübner)	variegated cutworm
Plathypena scabra (F.)	green cloverworm
Plodia interpunctella (Hübner)	Indian-meal moth
Prodenia ornithogalli (Guenée)	yellow-striped armyworm
Pseudaletia unipuncta (Haworth)	armyworm
Rhyacionia frustrana (Comstock)	Nantucket pine tip moth
Scotogramma trifolii (Rott.)	clover cutworm
Spilonota ocellana Denis & Schiffermüller)	eye-spotted bud moth
Spodoptera frugiperda (J. E. Smith)	fall armyworm
Spodoptera exigua (Hübner)	beet armyworm
Strymon melinus (Hübner)	cotton square borer
Trichoplusia ni (Hübner)	cabbage looper
Udea rubigalis (Guenée)	greenhouse leaf tier

3.122 Using Table 3.5, which order of insects is most responsive to blacklight traps?

- - - - - - - - - - - - - - - - -

Lepidoptera

Innumerable light sources have been used to observe insect attraction. These include the incandescent bulb, using a metal filament, and various gaseous discharge sources which employ mercury and other gases, such as argon, neon, and xenon. The fluorescent lamp, from which most of the ultraviolet (UV) or blacklight (BL) traps are made, is a form of the mercury-vapor discharge source. Mercury and argon lights have been the most used sources of UV light in recent studies of insect attraction.

UV light traps have come into widespread use in recent years as a survey device and have been extremely helpful in some insect-control programs. They are used extensively for detection and quarantine work and have the potential of being used as a control device for some economic pests. Some of the large field experiments and their results are shown in Table 3.6.

Insect light traps utilizing UV light may be used in six principal ways: (1) to detect the presence of imported noxious insects at ports of entry (detection traps); (2) to determine the spread and range of new pests in a region (survey); (3) to determine the seasonal appearance and insect abundance in a locality (survey); (4) to evaluate the effectiveness of control measures; (5) to control insects by reducing the population below the economic level; and (6) to supplement other control measures.

Advantages of light traps as a tool in IPM include avoiding insecticide residues on crops, allowing continuous operation during all weather conditions, allowing beneficial insects to play an important role in suppressing insect pests, and assisting in determining pest infestations without extensive field sampling.

On the other side of the coin, since not all species of insects are phototactic, light traps are not a panacea for every injurious insect. Evidence is accumulating, however, to suggest that they are finding their place in a variety of situations.

Generally, light traps have not been recommended for control purposes, although they are used on small acreages of high-priced crops, around factories where night-flying insects must be controlled, at outdoor fruit stands, and around drive-in restaurants and theaters, golf driving ranges, and dairy stables and milk houses.

The efficiency of UV traps is rather low. The only detailed research into trapping efficiency indicated 10 to 50 percent for cotton bollworm moths and 8 to 38 percent for cabbage loopers. Large numbers of attracted insects can be found very much alive around the traps during daylight.

3.123 Despite the large numbers of insects caught in blacklight traps, their actual efficiency is __________ .

- - - - - - - - - - - - - - - -

low

The distances that insects will travel to BL traps is not well known. Pink bollworm moths have been attracted experimentally 140 feet; corn earworm moths, 200 feet; and tobacco hornworm moths, 390 feet. It is likely that actual trapping ranges are much greater.

Virgin females or sex pheromones have been used very successfully in combination with UV lights to trap the

Table 3.6 Major large-scale UV light trapping projects of economic insects.

Location and crop	Trap type	No. traps	Test acreage	Insects involved and results
Red Rock, Arizona Lettuce, cotton	1 X 15W BL + CL pheromone	415	2000	Cabbage looper population reduced less than 50%. Economically unsuccessful. Terminated in 1970.
Oxford, N. Carolina Field tobacco	1 X 15W BL	340	113 sq. mi.	Hornworm population reduced an estimated 20%; with stalk destruction gave 55-94% hornworm reduction
Quincy, Florida Shade tobacco	2 X 15W BL + systemic + scouting	Spaced 160' apart on field borders	1.5 traps/acre	Tobacco budworm and hornworm. Insecticide applications reduced by 50%.
Quincy, Florida Field tobacco	1 X 15W BL + CL pheromone	1200	400 sq. mi.	Cabbage looper. Success not given.
Byron, Georgia Pecan trees	4 X 15W BL per acre	33	one orchard	Hickory shuck worm. Suppression equivalent to conventional applications of insecticide.
St. Croix, Virgin Islands No crop	1 X 15W BL + virgin females	250	84 sq. mi.	Tobacco hornworms. Population reduced to 14% after 3.5 years. No tobacco, only wild hosts.
Reeves County, Texas Cotton	1 X 15W BL and 1 X 32W incline 4 basic types	2000	16,000	Cotton bollworm and cabbage looper. Removed 19 bollworm moths/acre/day from population or reduction of 1 egg/200 stalks of cotton/day. Had little effect on cabbage looper population.

cabbage looper moth, tobacco budworm, pink bollworm, tobacco hornworm, and the Nantucket pine tip moth. Most nocturnal pest species will probably succumb to this irresistable combination.

3.124 Insect traps are most efficient when they combine ________________ with ____________________.

- - - - - - - - - - - - - - - - - -

blacklight pheromones

In 1965, a three-year study of the control of tobacco hornworms was conducted in North Carolina using BL traps in combination with stalk destruction immediately after harvest. Three traps per square mile were placed over a 113 square mile area. This combination resulted in a substantial reduction in hornworms within the trapped area and also reduced the need for insecticides. Some of the reduction was due, of course, to the stalk destruction. The use of a few traps over a small area is not expected to give favorable results. For tobacco growers to have a successful hornworm program using light traps, the following conditions were recommended: (1) the trapping area should be at least 12 miles in diameter (113 square miles), (2) light traps should be installed at the rate of three per square mile and distributed uniformly, and (3) all tobacco stalks should be destroyed immediately after harvest. (There were also indications of some control of the tobacco budworm on tobacco, however the control was not as complete as that of the hornworm.)

The codling moth is a key pest of apples. An outstanding study took place in Chile, in which the BL traps, using the small 4-watt lamps, were used to capture adults to determine the maximum emergence period and the best time to spray for first-brood larvae. The usual number of sprays to control the codling moth was reduced from three or four to two by precise timing based on light-trap catches.

It has been suggested by E. F. Knipling, developer of the sterile-male screwworm fly release, that a selective method of control which is only 50 percent effective against a target species may be more effective than a nonselective method that kills 90 percent of the target species and 90 percent of its natural enemies.

An example is the integrated program to control four major insects of cigar-wrapper tobacco in Florida. Treatments consisted of (1) a preplanting application of a soil systemic insecticide to control the tobacco flea beetle and aphids; (2) perimeter-mounted BL traps for tobacco budworm and cabbage looper; (3) applications of the insect pathogen *Bacillus thuringiensis*, methyl parathion, or azinphosmethyl. In the first year only two applications of insecticide were needed in comparison with seventeen for the check field. During the second year 75 percent fewer insecticide applications were needed for integrated fields as compared with check fields.

Despite the sensational catches of insects in UV light traps, they are not likely to become the single control method for any insect. We might conclude by saying that in the late 1970's light traps used in agriculture for the control of nocturnal insect pests have a position which belongs to one of supplemental efforts, as in monitoring and surveying with other control measures. These would be in conjunction with microbial control, (for example, *Bacillus thuringiensis*), cultural control (early crop plow-under for pink bollworm suppression), and chemical control with the use of insecticides after the insect pest reaches economic injury levels.

3.125 In reflecting on the use of blacklight traps and all of their possible uses, the most common long-range prediction for their use consists of an aid in ______________________________ insect pest populations.

- - - - - - - - - - - - - - - -

surveying, monitoring, and sampling

References

Barnes, M. M. et al. 1969. Reduction of treatment for codling moth in Chilean apple orchards by indexing with portable light traps. J. Econ. Entomol. 62: 733-734.

Cantelo, W. W. 1974. Blacklight traps as control agents: an appraisal. Bull. Entomol. Soc. Amer. 20(4): 279-282.

Gentry, C. R., W. W. Thomas, and J. M. Stanley. 1969. Integrated control as an improved means of reducing populations of tobacco pests. J. Econ. Entomol. 62(6):1274-1277.

Lam, J. J., Jr., A. H. Baumhover, C. M. Knott. 1970. Hornworm population suppression in a large area with traps using blacklight lamps. Paper 70-352, Ann. Meet. ASAE, Minneapolis, Minn., Jan. 7-10.

Sparks, A. N., R. L. Wright, and J. P. Hollingsworth. 1967. Evaluation of designs and installations of electric insect traps to collect bollworm moths in Reeves County, Texas. J. Econ. Entomol. 60(4): 929-936.

Taylor, J. G., L. B. Altman, J. P. Hollingsworth, and J. M. Stanley. 1956. Electric insect traps for survey purposes. USDA ARS 42-3. 11 pp.

USDA. 1965. Use of light traps supplemented with late season stalk cutting for control of hornworms on tobacco. ARS mimeograph. 3 pp.

Wolf, W. W., J. G. Hartsock, J. H. Ford, T. J. Henneberry, O. A. Hills, and J. W. Debolt. 1969. Combined use of sex pheromone and electric traps for cabbage looper control. Trans. ASAE 12(3):329-335.

3-F REGULATORY CONTROL

Fundamental regulatory control principles involve preventing the entry and establishment of foreign plant and animal pests in a country or area and eradicating, containing, or suppressing pests already established in limited areas. With the objectives of protecting the economy and welfare of the people, quarantine action is used to exclude potential pests, to prevent further spread of those already present, and to supplement eradication programs.

Ports of entry are the first line of defense against the introduction of new pests. Pests which break through this first line of defense are eradicated or contained within limited areas when possible. Quarantine action is used only against insects of economic importance, although it is sometimes necessary to contain insects which are of no economic importance in another country until their behavior in a new environment can be studied.

3.126 What are the fundamental regulatory control principles?

- - - - - - - - - - - - - - - -

To prevent the entry and establishment of foreign plant and animal pests and to eradicate, contain, or suppress pests already established in limited areas.

Legal Authority and Responsibility

The United States Department of Agriculture is responsible for regulatory control in this country. Authority for federal plant quarantine is presented in the Insect Pest Act of 1905, the Plant Quarantine Act of 1912, and several additional acts and amendments to these initial regulations.

Each state is responsible for enforcing quarantine regulations to prevent the spread of pests within the state. In addition, some states have regulations to prevent the introduction of insects from other states or countries. State Departments of Agriculture or similar agencies are usually responsible for enforcing the provisions of their states' regulations.

3.127 What federal and state agencies are responsible for enforcing quarantine regulations?

- - - - - - - - - - - - - - -

The United States Department of Agriculture and State Departments of Agriculture or similar state agencies.

Regulatory Control Programs

Three types of regulatory control programs are eradication, containment, and suppression. Although eradication of an insect species has never been accomplished, it is possible in some cases to eliminate a pest from a limited area. Examples of successful eradication of insects include elimination of the screwworm from the southeastern United States by use of a sterile male technique, and the Mediterranean fruit fly from citrus-growing areas by use of poison-bait sprays. Reinfestation is a problem in limited-area eradication efforts and often causes the necessity for repeated efforts against an insect.

Containment programs are used to limit the spread of insects that are likely to infest a larger area. Control of the gypsy moth along the edges of infestations in forests is a good example. This has helped to slow movement to the west and south from the northeastern United States where this pest was originally discovered.

Suppression programs are used when sudden outbreaks of insects occur over large areas and cannot be successfully dealt with by individuals alone. Outbreaks of grasshoppers in western states is an example of an insect problem which is controlled in this way. Federal and state regulatory agencies usually share the financial expenses of a program of this nature with the ranchers involved.

3.128 The three types of regulatory control programs are ________________, ________________, and ________________________.

- - - - - - - - - - - - - - -

eradication containment suppression

Importance of Regulatory Action in Insect Pest Management

Much of the value of regulatory action to IPM is indirect, and yet it should be recognized. Exclusion of a potentially destructive insect by preventing it's introduction into the United States certainly makes the IPM effort easier than if that pest were present. In a similar manner, preventing the spread of an insect within the United States is of benefit to IPM. It is important, however, that any regulatory action to eradicate, contain, or suppress an insect is based on sufficient information and technology to assure success of the action with minimal secondary problems.

Regulatory programs of state agencies can play vital roles in IPM. The value of such programs is somewhat related to the authority and responsibility of the state agency and its organizational quality. For example, in Texas, IPM in cotton has benefited greatly from a state regulatory program requiring cotton stalk destruction and plow-down at a date early enough to reduce overwintering populations of the pink bollworm and thus eliminate it as a major pest.

3.129 How can regulatory actions benefit IPM?

- - - - - - - - - - - - - - - -

1. By preventing the introduction of pests.
2. By preventing the spread of established insects.
3. By enforcing cultural and other control methods.

References

Anonymous. 1969. Insect-pest management and control. Nat. Acad. Sci. Publ. 1695.

Anonymous. 1952. Insects: The yearbook of agriculture. U.S. Department of Agriculture. 780 pp.

UNIT 4 POTENTIAL COMPONENTS OF THE INSECT-PEST-MANAGEMENT SYSTEM

We have deliberately selected the previous components of IPM—natural, cultural, biological, chemical controls—because they are known, proven entities in practice today. They are not enterprises requiring several more years of research before they can be practically utilized. The improvement in insect control utilizing these components, as well as other agricultural production practices, has resulted in greater land-use efficiency. Figure 4.1 shows that the carrying capacity of our land has almost doubled in the past twenty-five years, and Figure 4.2 shows that each farmer is now supplying farm products for approximately three times as many people as he did twenty-five years ago. Future improvements in all areas of crop production should result in a continual rise of the curves shown in these two figures. What we are about to discuss, however, is not ready for practical field application. Use of pathogens, pheromones, chemosterilants, and insect growth regulators are all relatively new and very much in the trial-and-error stage of laboratory-to-field transition.

There is an adage of scientists which, paraphrased, goes: Applied research grows out of the theories and discoveries of basic research. We have completed the truly practical aspects of IPM. The ditch bank, turnrow, nitty-gritty facets of IPM are covered. We now need to turn to the future and examine some of the exciting activity that is grasping the imagination of research scientists in experimental field plots and the laboratory.

The public is ready for innovations in pest control. One almost senses that they demand new, less disruptive techniques in managing pests. By the time news reporters translate scientific research for the average eighth-grade reader, it is difficult to determine just what was said. But eventually the right message will be delivered, and a new day dawns for the reader, the writer, and pest control.

Those new turns lie in the minds of thinkers, men who have imagination backed with solid scientific training, who can inspire others to produce the building materials needed for synthesis, the blending or cross-breeding of older ideas.

In 1972, as was mentioned earlier, a major step was taken to establish federal supervision of all pest control, the enactment of the FEPCA. This was a direct result of the President's Science Advisory Committee Report entitled Use of Pesticides (1963), the special study and report by the U.S. House of Representatives Committee on Appropriations entitled Effects, Uses, Control and

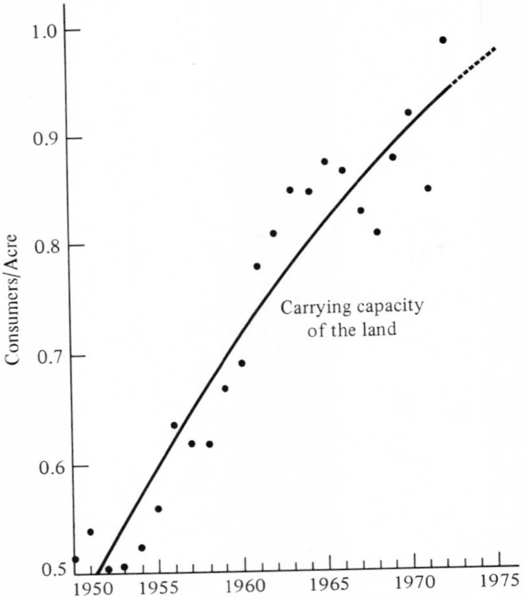

Figure 4.1 People supplied food per cropland acre, 1950–1972. (From Agricultural Production Efficiency, National Academy of Sciences, 1975.)

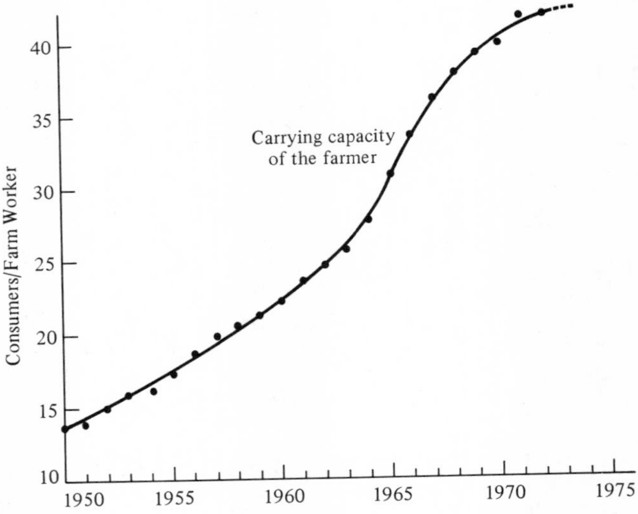

Figure 4.2 People supplied farm products per farmer, 1950–1972. (From Agricultural Production Efficiency, National Academy of Sciences, 1975.)

Research of Agricultural Pesticides (1966), led by former Congressman Jamie L. Whitten, and the report issued by the President's Science Advisory Committee entitled Restoring the Quality of our Environment (1965). These studies greatly influenced the final version of the FEPCA, and they will continue to influence future new legislation in the form of further amendments to the FIFRA of 1947.

Many activities involved in the overall control of insect pests have not emerged. These are possibly "itches that can't be scratched" in the minds of our thinkers, our synthesists. The discussions of the four following topics represent the fruition of thinking that began after the Civil War in the United States, with the introduction of arsenicals as stomach poisons for insects. This was the beginning of the entomologist's dream of manipulating the environment of the insect to the insect's detriment. Prior to the eventful insect discoveries, we could only adopt a fatalistic philosophy—what will be will be!

Now let us glimpse into the future!

4-A MICROBIAL CONTROL

Microbial control refers to the control of insect pest infestations by disease organisms. In one sense, it is inappropriately placed in the "potential" section of this manual. It more correctly belongs in a transitory section, indicating its current limited use as well as its vast potential for the future. Insect pathology is a relatively new field even though the idea has germinated slowly over hundreds of years. It was only after the establishment of the Laboratory of Insect Pathology at the University of California in 1945 that this discipline really gained momentum. The leadership and contributions of the late E. A. Steinhaus in this field make him the uncontested father of modern insect pathology and microbial control.

Microbial control as a potential tool in IPM programs owes allegiance to two currently used methods, chemical and biological controls. Some of the advantages and disadvantages of both methods apply also to microbial control. For example, some pathogens can be mass-produced (as are chemical insecticides in chemical control), applied in a conventional manner at certain dosage levels to kill an existing infestation, and dissipated in the environment. In such cases, the microbial agent is essentially a "living insecticide," and no prolonged or residual effects of the application are expected.

4.1 How are microbial agents similar to chemical insecticides?

\- - - - - - - - - - - - - - - - -

\- - - - - - - - - - - - - - - -

1. They are mass-produced.
2. They are applied in a conventional manner.
3. They are applied against existing pest populations.

Since microbial agents are living organisms, many of the principles that apply to other biological control agents, such as parasites and predators, apply equally as well to pathogens; for example, they may be introduced into an environment to initiate a disease outbreak. But the main effects of the pathogen come from reproduction and spread of the disease organisms in the pest population. In other words, as with parasites and predators, they are self-perpetuating and regulatory in nature. They remain in the environment and become a permanent mortality factor in the pest population. Good examples of this are the milky (spore) disease bacteria *Bacillus popilliae* Dutky and *B. lentimorbus* Dutky, used to control the Japanese beetle.

4.2 How are microbial agents similar to other biological control agents?

\- - - - - - - - - - - - - - - -

1. They are living organisms.
2. Many persist in the environment and continue to act on pest populations.
3. They are regulatory in nature.

The bacterium *Bacillus thuringiensis* Berliner is a pathogen used for insect control in a manner similar to the use of conventional insecticides. It is the only microbial agent now registered for use on food crops in the United States. Since it is short-lived, repeated applications are necessary.

4.3 What insect pathogen is now used for control of certain pests?

\- - - - - - - - - - - - - - - -

Bacillus thuringiensis

Pathogenic bacteria, viruses, fungi, nematodes, and protozoa affect a wide range of insects, beneficials and pests alike. In nature, they play a large role in regulating insect pest numbers. In most agroecosystems practically no year passes without almost complete decimation of cabbage loopers by a naturally occurring nuclear polyhedrosis virus. Similarly, a polyhedrosis virus of the alfalfa caterpillar is important in the natural control of this pest. Under favorable humidity and temperature conditions, pathogenic fungi play an important part in the natural control of a wide variety of insects. For example, several species of fungi are parasitic on the spotted alfalfa aphid.

Although many insect pests are subject to mortality from pathogenic agents occurring naturally in the environment, little reliance can be placed on them because of their unpredictable nature. Much research has been done in an effort to understand better the relationship of the three important components of a disease outbreak: the host insect, the pathogen, and the environment. As information is accumulated on these organisms and as a better understanding of their ecological requirements is gained, the importance of microbial control in IPM programs will surely increase to the point of their becoming a major tool in the total management scheme. For example, some pathogens exhibit high virulence against certain pests in the laboratory, but under field conditions relatively little effect is observed. The virus of the corn earworm is a case in point. First attempts to utilize the nuclear polyhedrosis virus for field control of this pest were totally unsatisfactory because ultraviolet radiation rendered the virus ineffective. Several formulations of this virus have been prepared in an attempt to shield the virus particles from excessive radiation, and these have increased effectiveness under field conditions in Arizona (Figure 4.3).

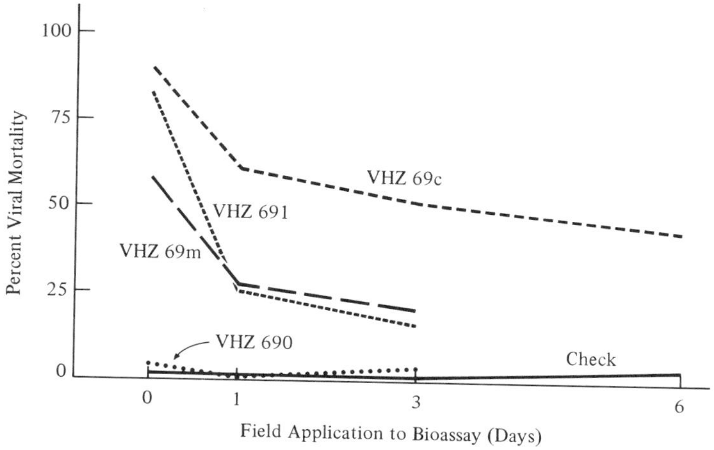

Figure 4.3 Bioassay determinations of VHZ persistence at four intervals following field application.

4.4 Why cannot disease outbreaks generally be relied upon in insect control?

- - - - - - - - - - - - - - -

Because of their unpredictable nature. Reliance will require a better understanding of the interrelationship of the host insect, the pathogen, and the environment.

Although the use of *B. thuringiensis* for control of pest insects is beyond the "potential" stage, its full utilization lies ahead. The "potential" of the many other promising pathogens, particularly the viruses, also offers much hope for adding another important weapon to the IPM arsenal. This is true whether the pathogens more nearly resemble the conventional insecticides (as in chemical control) or the beneficial insects (as in biological control) in terms of their effects on pest populations.

References

Falcon, L. A. 1971. Microbial control as a tool in integrated control programs. Pp. 346-364 in Biological control, ed. C. B. Huffaker. Plenum Press, New York-London.

Falcon, L. A. 1974. Insect pathogens: integration into a pest management system. Pp. 618-627 in Proceedings of the Summer Institute on Biological Control of Plant Insects and Diseases, ed F. G. Maxwell and F. A. Harris. University Press of Mississippi State University, Starkville, Miss.

Hall, I. M. and P. H. Dunn. 1957. Entomophthorous fungi parasitic on the spotted alfalfa aphid. Hilgardia 27:159-181.

Steinhaus, E. A. 1949. Principles of insect pathology. McGraw-Hill, New York. 757 pp.

Steinhaus, E. A. 1956. Microbial control—the emergence of an idea: A brief history of insect pathology through the nineteenth century. Hilgardia 26:107-160.

Steinhaus, E. A. 1964. Microbial diseases of insects. Ch. 18 in Biological control of insect pests and weeds, ed. P. DeBach. Reinhold Publishing Corporation, New York. 844 pp.

Steinhaus, E. A. and C. G. Thompson. 1949. Preliminary field tests using a polyhedrosis virus to control the alfalfa caterpillar. J. Econ. Entomol. 42: 301-305.

White, R. T. and S. R. Dutky. 1940. Effect of the introduction of milky diseases on populations of Japanese beetle larvae. J. Econ. Entomol. 33(2):306-309.

4-B INSECT PHEROMONES

Many insects communicate with chemicals. They release very small quantities of highly specific compounds that vaporize readily and are detected by insects of the same species. As we learn more about insect behavior, we may discover that all insects communicate through the release of these chemical agents known as pheromones.

Insect pheromones are probably the most potent physiologically active components known today. Pheromone comes from the Greek "pherein," to carry, and "horman," to excite or stimulate. Pheromones are excreted to the outside of the insect body, where they cause specific reactions from other insects of the same species; they are also referred to in older literature as "social hormones."

4.5 The word pheromone means ____________ of ____________.

\- - - - - - - - - - - - - - - -

carrier excitement

Based on the behavioral response of the receiving insect, pheromones can be classified into the following behavioral categories: sexual behavior, aggregation (including trail following), dispersion, oviposition, alarm, and specialized colonial behavior.

Of the different types of pheromones, it is the sex pheromones that offer the greatest potential to IPM systems. A good example of their recent use took place in the Texas high plains boll-weevil-diapause control zone. Pheromone traps using live male boll weevils captured sufficient overwintered weevils to suppress the population until late summer migration overpowered the action of the traps.

4.6 Of the different types of pheromones, the ____________ pheromones offer the greatest potential for use in IPM systems.

\- - - - - - - - - - - - - - - -

sex

Sex pheromones of Lepidoptera (moths and butterflies) have received the most detailed chemical study to date. For instance, after thirty years of study, the gypsy-moth sex pheromone was isolated, identified, and synthesized in the laboratory in 1960. Since then great quantities of Disparlure, the synthetic female gypsy-moth sex pheromone, have been used in male-trapping programs for this forest pest.

A current list of the available synthetic sex pheromones is presented in Table 4.1.

Table 4.1 Some of the commercially available synthetic insect sex pheromones

Common name	Trade name	Species attracted*
disparlure	Disparmone® Pherocon GM	gypsy moth
grandlure	Grandamone® Pherocon BW	boll weevil
gossyplure	Pherocon PBW	pink bollworm
hexalure**	Hexamone®	pink bollworm
looplure	Cabblemone® Pherocon CL	cabbage looper
muscalure	Muscamone®	house fly
codlelure	Codlemone®	codling moth
virelure	none	tobacco budworm
none	Z-11	red-banded leaf roller European corn borer oblique-banded leaf roller smartweed borer

* See Appendix for scientific names
** Obsolete

There are no current citations of complete insect control using synthetic pheromones, but many species are under investigation. The uses of synthetic pheromones include attracting mate-seeking insects to mechanical or sticky traps, to insecticide-treated areas, to poisoned baits, or to ultraviolet light traps. A method under study is designed to distribute gossyplure (the pink bollworm sex pheromone) in cotton fields in the spring to confuse emerging males and prevent mating. This principle could conceivably be applied to most insects when synthetic pheromones are available in large quantities.

The uses of sex pheromones in IPM systems seems almost endless. A few of these potential applications include

1. Using pheromones in a sterilization program of native insects to attract the insects to the chemosterilant source;
2. Improving the mating competitiveness of released sterilized insects by making them more attractive to native populations;

3. Using pheromones to concentrate pests in limited areas for treatment with an insecticide or cultural method;
4. Using relatively large quantities of pheromones in selected areas for male confusion;
5. Using pheromone traps to capture the few overwintering or emerging adults in the spring as a means of early population suppression.

Food and oviposition attractants are required in much greater quantities than sex pheromones and are considered as lures rather than pheromones. In Table 4.2 are shown some of the synthetic insect attractants, which are essentially feeding lures. The Japanese beetle has long been trapped in the eastern United States by using a fragrant food attractant, geraniol. Fly traps have been developed over the years that use oviposition attractants in the form of ammonia-yielding chemicals or decaying organic matter.

Table 4.2 Some synthetic attractants for insects (Beroza, 1970).

Common name	Trade name	Species attracted*
trimedlure	Pherocon MFF	Mediterranean fruit fly
cue-lure	Q-lure Pherocon QFF	melon fly
methyl eugenol	None	Oriental fruit fly
2,4-hexadienyl butyrate	None	yellow jackets
heptyl butyrate	None	yellow jackets
amlure	None	European chafer
methyl cyclohexanepropionate and eugenol	None	Japanese beetle
rhinolure	None	Rhinoceros beetle

* See Appendix for scientific names.

4.7 If we consider pheromones one class of attractants, two other classes would be ____________ and ____________ lures.

\- - - - - - - - - - - - - - -

food oviposition

Despite the exuberance with which the potentials of sex pheromones have been praised, pheromones are most practically used in survey traps to provide information about population levels, to delineate infestations, to monitor control or eradication programs, and to warn of new pest introductions. It is likely that IPM systems of the future will rely heavily on the two key uses of pheromones, survey and removal of early-emerging populations through trapping.

4.8 The use of pheromones in insect control is still rather limited to ________________ and ________________ insect populations.

- - - - - - - - - - - - - - - -

surveying monitoring

References

Shorey, H. H. 1973. Behavioral responses to insect pheromones. Ann. Rev. Entomol. 18:349-380.

Beroza, Morton, ed. 1970. Chemicals controlling insect behavior. Academic Press, New York. 170 pp.

New Approaches to Pest Control and Eradication. 1963. Symposium of the American Chemical Society, Atlantic City, N.J. 11 Sept. 1962. Symp. Chmn., S. A. Hall. Advances In Chemistry Series 41. American Chemical Society, Washington, D.C.

Proceedings of the Tall Timbers Conference on Ecological Animal Control by Habitat Management. 1974. Tallahassee, Fla., 1-2 Mar. 1973. Conf. Chmn., E. V. Komarek, Sr. Tall Timbers Res. Sta., Tallahassee, Fla. 178 pp.

4-C CHEMOSTERILANTS

Chemosterilants are defined as chemical compounds which are capable of depriving insects of their ability to reproduce. Over one thousand compounds that affect reproduction in insects have been described. These come under the very broad classification of chemosterilants. The massive research effort to uncover these sterilizing chemicals is a direct spin-off of the success of programs for the eradication of the screwworm by the release of males sterilized by gamma radiation. This concept involves the use of insects for the destruction of their own species through the induction of sterility in a large proportion of the males. It utilizes their mating behavior to reach female insects that would not be affected by the usual insecticidal control techniques.

4.9 Chemosterilants are chemicals which prevent insects from __________________.

\- - - - - - - - - - - - - - - -

reproducing

The advantages of using chemicals instead of rearing, irradiating, and releasing massive numbers of males to induce sterility in a population are obvious. When it was shown that both males and females could be sterilized simultaneously, the potential seemed to be limitless. The development of chemosterilants is relatively new, and no definitive statements can be made about all types of compounds with sterilizing activity.

4.10 What would be the obvious advantages of using chemosterilants over rearing, irradiating, and releasing sterile insects to sterilize a large population of insects?

\- - - - - - - - - - - - - - - -

Using chemosterilants is easier, is less expensive, requires less labor, and is faster.

There are essentially four classes of chemosterilants: alkylating agents, phosphorus amides, triazines, and antimetabolites.

Alkylating Agents

Alkylating agents constitute the largest and most active class of chemosterilants. They are moderately to highly reactive compounds with proteins and nucleic acids. They are sometimes referred to as radiomimetics (radiation mimicking materials) in that their effects are similar to these of X-or gamma-rays. These agents replace hydrogen in fundamental genetic material with an alkyl group ($-CH_3$ or $-C_2H_5$, and so on), which results in an effect similar to irradiation. They are highly effective in producing mitotic disturbances or nucleotoxic conditions, particularly in tissues where cell division and multiplication take place at a high rate. This results in the production of multiple dominant lethal mutations or severely injured genetic material in the sperm or the egg. Although fully alive, zygotes (fertilized eggs) if formed do not complete development into mature progeny.

4.11 Alkylating agents react with ________________ and ________________ by substituting an ____________ group for ________________ in these molecules.

- - - - - - - - - - - - - - - -

proteins   nucleic acids   alkyl group   hydrogen

4.12 These chemosterilants damage genetic material in the ______________ or the ______________.

- - - - - - - - - - - - - - - -

sperm   egg

The two most widely investigated types of alkylating chemosterilants are aziridines and alkanesulfonates. The chemical and physical properties of these compounds are quite variable, but their cytotoxic and mutagenic effects are closely related. Although the alkylating agents are relatively unstable and degrade rapidly, the possible contamination of large areas, even with small residues, makes their use as crop sprays or dusts hazardous and undesirable. Safe applications are possible, however, when these chemosterilants are used to sterilize reared or collected insects under controlled conditions and when personnel are adequately protected. Alkylating agents shown in Table 4.3 are apholate, tepa, metepa, thiotepa, tretamine, and bisulfan.

Phosphorus Amides

Some of the most effective aziridinyl chemosterilants (tepa and thiotepa) are phosphorus amides, but are classified as alkylating agents because of their high alkylating reactivity. The phosphorus amides do not contain alkylating functional groups and do not react directly with nucleophilic receptors. Hempa and thiohempa belong to this group and are cytotoxic and mutagenic, but at much higher dosages than the alkylating agents. Their residues are longer lasting than those of the alkylating agents, making their use as crop sprays or dusts even less desirable.

Triazines

The triazine chemosterilants are similar to the herbicidal triazines, and for this reason their use will be substantially limited when broad-leafed plants can be contaminated. The triazines are mutagenic and cytotoxic and have somewhat longer residual action than the other classes. Hemel is an example of the triazines (Table 4.3).

Table 4.3 Common and chemical names, mammalian oral LD_{50}s, and chemical structures of the more common chemosterilants.

Common and chemical names	Oral LD_{50} mg/kg	Chemical structure
Apholate 2,2,4,4,6,6-hexakis(1-aziridinyl)-2,2,4,4,6,6-hexahydro-1,3,5,2,4,6-triazatriphosphorine	98	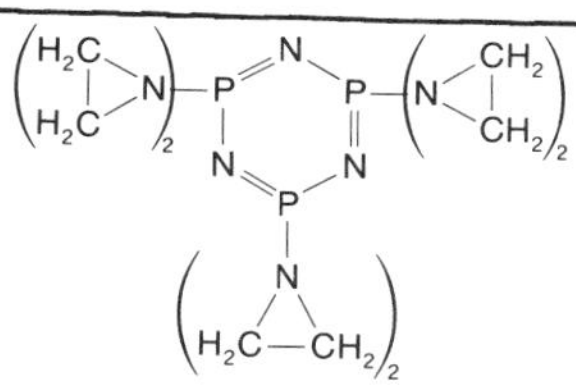
Tepa tris(1-aziridinyl)phosphine oxide	37	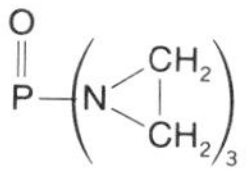
Metepa tris(2-methyl-1-aziridinyl)=phosphine oxide	136	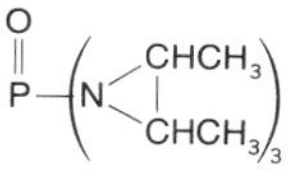
Thiotepa tris(1-aziridinyl)phosphine sulfide	9	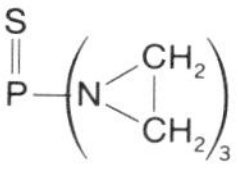
Tretamine 2,4,6,-tris(1-aziridinyl)=*s*-triazine	1	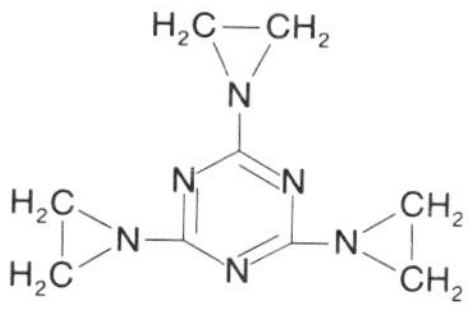
Busulfan 1,4-butanediol dimethane=sulfonate	18	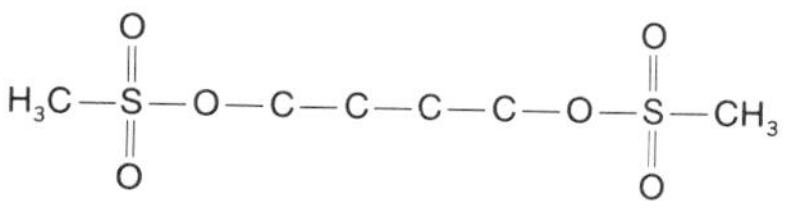
Hempa hexamethylphosphoric triamide	2,650 ♂ 3,360 ♀	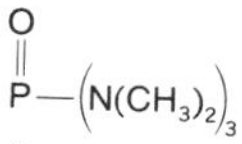
Thiohempa hexamethylphosphorothioic triamide	20	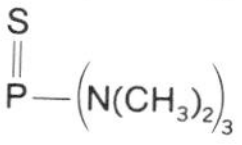
Hemel hexamethylmelamine	350	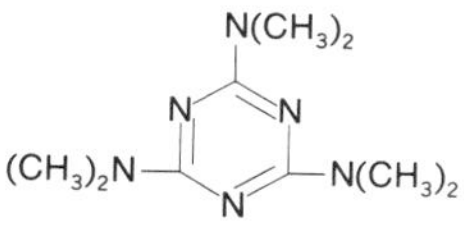

Antimetabolites

Antimetabolites are compounds that are chemically and structurally similar to important biochemicals that occur naturally in an insect; they replace or displace these biochemicals, rendering the metabolic process useless if not toxic. Because the concept of antimetabolites has not been used on chemically or structurally similar compounds, no generalizations about their properties can be made. At this time the specific antimetabolites are only of research interest with no immediate field potential.

Field Use

The alkylating chemosterilants have been used in the field with moderate to good success for experimental house fly control around garbage and trash dumps. Busulfan fed to boll weevils which were later released in the field resulted in only moderate success. Numerous laboratory studies indicate great potential for the chemosterilant principle, but the obvious problem is the hazardous nature of the residues on food and feed crops. Again, we are dealing here with another form of chemical control, where modes of action differ from those of conventional insecticides.

4.13 Are the hazards of spraying chemosterilants on crops any different from those of spraying one of the common insecticides?

- - - - - - - - - - - - - - - - -

Yes. We are spraying materials which can injure genetic material in all animals, including humans. Their toxicity is not so much to the parent but to the future offspring.

4.14 What are the four classes of chemosterilants?

- - - - - - - - - - - - - - - - -

Alkylating agents, phosphorus amides, triazines, and antimetabolites.

References

Borkovec, A. B. 1972. Safe handling of insect chemosterilants in research and field use. ARS, USDA. ARS-NE-2 (November).

Alexander, P. 1960. Radiation-Imitating Chemicals. Sci. Amer. 202(1):99-108.

LaBrecque, G. C. and C. N. Smith, eds. 1968. Principles of insect chemosterilization. Appleton-Century-Crofts, New York. 354 pp.

4-D INSECT GROWTH REGULATORS

Insect growth regulators (IGRs) is a term encompassing a relatively new group of chemical compounds that alter growth and development in insects. Their effects have been observed on embryonic and larval and nymphal development, on metamorphosis, on reproduction in both males and females, on behavior, and on several forms of diapause. They include ecdysone (the molting hormone), juvenile hormone (JH), JH mimic, JH analog (JHA), and their broader synonyms, juvenoids and juvegens.

A more common term for the IGRs is "third-generation" insecticides. First-generation insecticides are the stomach poisons, such as the arsenicals; the second-generation includes the familiar organochlorine, organophosphate, carbamate, and formamidine insecticides. The third-generation insecticides, then, are the ultimate in selectivity, affecting only specific insects, and apparently having no undesirable effects on humans, wildlife, and the environment. Consequently, they are highly compatible with IPM principles.

4.15 Why are the IGRs so very compatible with the principles of IPM?

- - - - - - - - - - - - - - - -

They affect only insects. They have no biomagnification, no persistence, and no undesirable effects on other animals.

Several glands in insects are known to produce hormones, the principal functions of which are the control of reproductive processes, molting, and metamorphosis. Here we are interested only in the hormone ecdysone, which is responsible for molting, and JH, which inhibits or prevents metamorphosis.

4.16 The hormone responsible for molting in insects is ________________.

- - - - - - - - - - - - - - - - -

ecdysone

Applications of ecdysones to insects are usually lethal in all stages of growth, making ecdysones similar to second-generation insecticides. One attractive feature of ecdysones as potential tools is their widespread distribution in plants. More than forty compounds have been isolated from higher plants, and these may play as yet unrecognized roles in insect-plant relationships.

Keen interest has been directed toward JHs. These are not, in the usual sense, toxic to insects. Instead of killing directly they interfere in the normal mechanisms of development and cause the insects to die before reaching the adult stage.

Dramatic results have been obtained in the laboratory, the most promising effects being on mosquite larvae, caterpillars, and certain beetle larvae, although effects have been observed on practically all insect orders. Most insect species respond to treatment with JHs by producing extra larval, nymphal, or pupal forms that vary from giant, almost perfect forms to intermediates of all sorts between the immatures and the adults. For the most part, the periods of greatest sensitivity for metamorphic inhibition are the last larval or nymphal instars, and the pupa in those having complete metamorphosis. One recognizable problem is the precision of timing applications to achieve maximum damaging effect on the upcoming life stage of a particular insect.

4.17 What is the "precision" problem in using JHs for insect control?

- - - - - - - - - - - - - - - - -

The need for precise timing to obtain greatest effect on immature stages.

For practical purposes, IGRs could be used on crops to suppress damaging insect numbers. They would be applied with the purpose of preventing pupal development or adult emergence, thus keeping the insects in the immature stages resulting eventually in their deaths (Table 4.4).

Table 4.4 Effect of a synthetic growth regulator on the cotton leafperforator in the greenhouse.

Treatment	Mean no. adults collected from caged plants over period of 33 days	Mean no. pupal cases inside cages at end of experiment
Untreated check	102.5	124.5
Foliar spray with IGR	1.5	1.0

Until recently, it was believed that insects could not develop resistance to JH, since it was an integral part of their physiology. However, this bubble, too, has burst, for certain strains of insects that show resistance to insecticides also show cross-resistance to insect JH mimics and JHAs.

4.18 Insecticide resistance has become a universal problem. Is there a likelihood that insects will develop resistance to IGRs?

- - - - - - - - - - - - - - - -

Yes, according to the same basic genetic principles as witnessed in insecticide resistance.

To date, the only IGR registered by the EPA is Altosid®, a mosquito growth regulator, manufactured by the Zoecon Corporation. Early in 1975 it was registered for use against second through fourth larval instar floodwater mosquitoes at 3 to 4 ounces per acre, to prevent adult emergence. Larvae exposed to Altosid continue their development to the pupal stage where they die. Altosid has no effect when applied to pupae or adult mosquitoes. Currently its use is limited to public health officials and other trained personnel of public mosquito abatement programs.

Can IGRs become successful pest-control agents? Certainly they can in time. They will, however, have to meet the general criteria for other pest-control agents; thus, they must be effective in reducing insect populations below economic damage levels, be competitive with second-generation insecticides in cost, and have no undesirable side effects. In summary, it appears that IGRs hold intriguing possibilities for future use in practical IPM systems. It should also be kept in mind that IGRs are chemicals and fall in the category of chemical control, but they have modes of action different from the traditional insecticides.

4.19 If third-generation insecticides are the IGRs, what are the first- and second-generation insecticides?

- - - - - - - - - - - - - - - -

First-generation insecticides are the stomach poisons; second-generation insecticides are the contact organochlorines and organophosphates.

References

Staal, G. B. 1975. Insect growth regulators with juvenile hormone activity. Ann. Rev. Entomol. 20:417-460.

Menn, J. J. and M. Beroza, eds. 1972. Insect juvenile hormones, chemistry and action. Academic Press, New York. 341 pp.

Williams, C. M. 1967. Third-generation Pesticides. Sci. Amer. 217(1):13-17.

Vinson, S. B. and F. W. Plapp, Jr. 1974. Third generation pesticides: The potential for the development of resistance by insects. J. Agr. Food Chem. 22(3): 356-360.

UNIT 5 *IMPLEMENTING PRACTICAL INSECT PEST MANAGEMENT*

The basic elements of IPM, emphasized in Unit 2, are natural control, sampling, economic levels, and insect biology and ecology. Of these, sampling is probably of most concern in establishing an IPM system. A sampling program to provide information on insect numbers in each field must be developed to serve as a base for utilizing knowledge of natural control, economic levels, and insect biology and ecology. Once the sampling program has been established, the four elements can be dovetailed together to serve as the foundation upon which practical components can be added to form the total IPM system. Because of the importance of establishing the sampling program, emphasis is placed on this basic element of the system. In addition, suggestions are offered for developing the total IPM system.

5.1 The basic element of most concern in establishing an IPM system is ____________________.

- - - - - - - - - - - - - - - -

sampling

5-A ESTABLISHMENT OF INSECT PEST MANAGEMENT PROGRAMS

The related personnel units involved in establishing an IPM program are a grower committee or organization, advisory personnel, supervisors, and field samplers, or "scouts." The grower organization is very important in establishing an IPM program. Conduct of the program by a grower committee avoids the stigma of being a government- or university-sponsored program and imposes the responsibility on the growers. Despite the use of federal funds to help initiate a program, the grower organization should maintain a leadership position with a definite plan for eventually assuming complete program responsibility. A committee or board, composed of seven or more growers, should be selected to head the organization. The committee is responsible for meeting with the advisory personnel and deciding on a course of action, making financial arrangements for conducting the program, hiring supervisors, organizing the growers, obtaining the necessary insurance on personnel and crop loss, and meeting regularly to discuss and evaluate the program. In arranging financing it may be advantageous for the growers to work with existing organizations, such as gin groups and grower cooperatives.

5.2 Why is the grower organization important in IPM?

- - - - - - - - - - - - - - - - -

It puts the growers in a responsible position and helps the growers to recognize the activity as their program.

Advisory personnel include extension and research entomologists, county agents, and other entomology professionals available to the organization. These persons are responsible for assisting the grower committee or board in organizing the program, training the field scouts in insect identification and techniques of field sampling, assisting with weekly meetings of scouts, supervisors, and growers to discuss problems and situations, and providing the growers with information on insects and their control by IPM.

5.3 What are the responsibilities of advisory personnel in establishing an IPM program?

- - - - - - - - - - - - - - - - -

1. To assist the growers in organizing the program.
2. To train the field scouts.
3. To assist with weekly meetings.
4. To provide information on insects and their control by IPM.

Supervisors are trained entomologists or persons experienced in conducting IPM programs. They are responsible for hiring the required field scouts, assigning scouts to the area and fields for which they are responsible, providing scouts with field record forms, discussing field record information with growers, rechecking fields with borderline infestations, and assisting growers in identifying and applying on a timely basis the practical IPM components available.

5.4 What are the qualifications of a supervisor?

- - - - - - - - - - - - - - - - -

To be trained entomologist or a person experienced in conducting IPM programs.

Field scouts are primarily college students, although they may be school teachers, high school students, and persons employed specifically for this purpose on a year-round basis. Usually they are hired locally and for the period when crop scouting is required. The field scout is responsible for properly sampling and recording information weekly, or as often as required, on insect conditions in assigned fields, providing completed record sheets to the grower or supervisor as arranged, and attending training sessions, including weekly meetings.

5.5 What are the duties of field scouts?

- - - - - - - - - - - - - - - -

1. To properly sample assigned fields and record the information.
2. To provide complete record sheets to the grower or supervisor.
3. To attend training sessions, including weekly meetings.

A few special preparations are essential in initiating a sampling program. It is necessary to obtain aerial photomaps that clearly show the fields to be sampled. These fields should be numbered for identification by means of a four-digit numbering system. This is needed to provide a reserve of potential numbers and to permit indefinite assignment of field numbers. The fields should be assigned to scouts in time for them to study both appropriate aerial and road maps to locate the fields prior to or during the first week of sampling.

5.6 Why is a four-digit numbering system advisable in field identification?

- - - - - - - - - - - - - - - -

It provides a reserve of potential numbers and permits leaving a field number unchanged indefinitely.

Ideally, scouts should work in pairs. Important advantages of working in pairs are efficiency in transportation and the availability of another person for discussion of problems or situations that may arise. Side benefits are the maintenance of integrity and avoidance of tedium.

5.7 What are two important advantages of scouts working in pairs?

- - - - - - - - - - - - - - - - -

1. Efficiency in transportation.
2. Availability of another person for discussion of problems or situations that may arise.

Good communications among scouts, supervisors, and growers is imperative in making an IPM program function smoothly. Procedures should be developed to insure frequent and continuous communication among the individuals throughout the season. Installation of telephones in vehicles used by supervisors, particularly where long distances or large acreages are involved, expedites communications and thus makes field assistance more timely.

To develop the IPM system, supervisors and advisory personnel must clearly identify and promote the use of the various practical components which are available to growers. Once the sampling program is established, the implementation of these components in an effective way becomes much easier. Some components, such as better use of beneficial insects and judicious use of insecticides, are utilized during the season and rely heavily on the sampling information collected by field scouts. Other components, such as the use of resistant varieties and certain cultural practices, must be incorporated at the appropriate time of the year and usually when there is no immediate threat from the insects they are meant to control. It is essential to emphasize the use of all practical components on a year-round basis. Chemical control should be used in a selective manner when, and only when, economic levels occur, despite the application of all known alternate components. As the IPM system matures in an area it should involve all crops grown. This permits consideration of the interrelationships of pest problems among crops and enables more complete utilization of the practical components of the IPM system.

5.8 How should chemical control be utilized as a component of the IPM system?

- - - - - - - - - - - - - - - -

In a selective manner and only when economic levels occur, despite the application of all known alternate components.

References

Carruth, L. A. and L. Moore. 1973. Cotton scouting and pesticide use in eastern Arizona. J. Econ. Entomol. 66:187-190.

Moore, L. 1972. The pest management system in cotton. Implementing practical pest management strategies: Proceedings of a National Extension Insect-Pest Management Workshop. Purdue University, West Lafayette, Ind. Mar. 14-16.

Watson, T. F. and L. Moore. 1970. Cotton insect control in Arizona—a changing outlook. Prog. Agr. Ariz. 22: 6-9.

5-B EXISTING INSECT PEST MANAGEMENT PROGRAMS

Theoretically all crops and their insect pests lend themselves readily to IPM. In practice this is not yet true. Those that do must meet the first two criteria for a functional program: realistic insect-pest sampling methods and economic injury levels for key pests. Those crops that have IPM programs currently in use, or those emerging from the tremendous amount of research required to place them on solid footing, will be discussed. They are cotton, corn, alfalfa, tobacco, grain sorghum, peanuts, citrus, vegetables, tree fruit, and soybeans. These ten are found in the top seventeen most important crops of the nation, presented in Table 5.1.

Cotton

Cotton IPM principles were developed by Dwight Isely in Arkansas in the late 1920's. That early system consisted of (1) establishing an economic level for the boll weevil, (2) weekly sampling of each field by cotton scouts, (3) applying insecticides only when necessary, (4) planting early-maturing varieties, and (5) early harvesting followed by destruction of crop residue. Although some of these principles have been continued and further developed in certain states, national emphasis was focused on chemical control beginning with the advent of the insecticide era in the mid-1940's.

It was not until 1971, when a pilot cotton IPM project was established in Pinal County, Arizona, that renewed emphasis was placed on the development of the cotton IPM system. This pilot program was initiated with the assistance of federal grant funds after dramatic cost-saving results were made known of a private, grower-financed program in Graham County, Arizona, begun in 1969 (Table 5.2). The national interest in environmental improvement at the time prompted approval of proposals to increase this kind of effort through federal funding.

Table 5.1 Acreages of some of the more important crops grown in the United States and their average yields per acre.[a]

Crop	Acreage 1974	Annual Production 1974	Yield/Acre (Normal)
Wheat	65,459,000	1,793,322,000 bu.	32 bu.
Corn	65,194,000[c]	4,651,167,000 bu.	92 bu.
Soybeans	52,460,000[c]	1,233,425,000 bu.	28 bu.
Alfalfa	26,642,000[c]	74,293,000 tons	2.85 tons
Grain sorghum	13,917,000[c]	628,081,000 bu.	59 bu.
Oats	13,325,000	620,539,000 bu	48 bu.
Cotton	12,669,600[c]	11,701,800 bales	510 lbs. lint
Barley	8,281,000	308,077,000 bu.	42 bu.
Vegetables	3,338,200[b,c]		
Rice	2,569,000	114,096,000 cwt.	44.4 cwt.
Dry beans & peas	1,782,700	24,033,000 cwt.	14.2 cwt.
Deciduous fruits	1,531,400[b,c]		
Peanuts	1,477,100[c]	3,679,963,000 lbs.	2490 lbs.
Potatoes	1,380,700	340,116,000 cwt.	246 cwt.
Sugar beets	1,216,500	22,268,000 tons	20 tons
Citrus	1,181,700[b,c]	13,393,605	
Tobacco	961,800[c]	1,958,214,000 lbs.	2000 lbs.
Sugar cane	747,500	25,760,000 tons	35 tons
Nut trees	396,400[b,c]		

[a]USDA 1975. Crop Production, 1974 Annual Summary. CrPr2-1(75). Crop Reporting Bd. Statistical Report. Serv. Washington, D.C.

[b]USDA 1974. Agricultural Statistics, 1974. (Data are for 1973.)

[c]Denotes those crops for which IPM projects are active or being developed through intensive field research.

Table 5.2 Effects of a cotton insect-pest-management program on practices and costs of insect control in Graham County, Arizona.

	1968*	1969	1970	1971
Program acres	13,263	12,750	9,655	11,051
Scouting cost/acre	------	1.67	1.63	1.65
Sprayed acres	13,263	2,040	1,738	3,190
% of total (1 or more appl.)	100	16	18	29
Avg. appl./acre	6	1.6	1.7	3.8
Cost/appl./acre	$2.76	$3.60	$3.62	$3.55
Total cost	$220,000	$11,750	$10,687	$43,069

*Before IPM program.

Following the pilot effort, cotton IPM projects were established in each of the fourteen major cotton-producing states in 1972. In that year, approximately 487,000 acres of cotton were scouted by four hundred Extension Service trained scouts in IPM projects, and these figures have increased steadily each year.

The basic objective of cotton IPM programs is to improve cotton-insect-control efficiency. The major elements which have been developed as a base for the IPM system are the scouting programs and the use of sound economic levels. Scouting programs, to provide good field sampling, were the first priority and have been successfully established in the project areas. Economic levels are available for the major cotton insects in different areas and are being improved by current research efforts.

Components of the cotton IPM systems vary from one area of the country to another because of the insects involved. Generally, however, they consist of the following:

Cultural control

Early harvest and plow-down are encouraged in most areas. These practices have been especially useful as a means of reducing pink bollworm populations. Practices such as irrigation termination in irrigated areas are being used to hasten crop maturity before large numbers of pink bollworm larvae are in diapause and capable of overwintering.

Proper cultivation, irrigation, and fertilization practices are encouraged to produce healthy, fruiting cotton plants, which helps to minimize insect damage. In addition to affecting fruit set, too much fertilizer or water increases problems with such insects as plant bugs and bollworms, while too little may be favorable to white flies and cotton leafperforators.

Strip-cutting or block-cutting alfalfa has been used to hold lygus bugs in this crop and avoid their migration into cotton, their normal behavior following solid cutting. Trap crops have also been used as a means of controlling bollworms and to some extent boll weevils. To control boll weevils, strips of early-maturing cotton are planted early to permit fruiting ahead of other cotton in the area. These strips are treated with a systemic insecticide, thus serving as a trap for boll weevils. This reduces populations which normally disperse over the entire area.

Biological control

The use of biological control has been greatly enhanced in IPM programs by avoiding use of insecticides until economic levels are reached. Predators and parasites often hold insects such as the cotton bollworm under control unless their own populations are reduced by insecticides. Permitting subeconomic insect numbers to exist in cotton helps to maintain the beneficial insect populations at a level capable of controlling many insect pest problems. This is encouraged through the economic level concept of IPM.

Introduction of parasites and predators is not yet widely used in cotton IPM, although there is some use of the small parasitic *Trichogramma* wasps for bollworm control and of lacewing larvae for control of bollworms and other insect pests. Considerable research is currently in progress on the introduction of beneficial insects, and this method may increase in importance in the near future.

Chemical control

Good field sampling and use of economic levels permit the proper selection and timing of insecticide applications on a need basis. A field is treated only when it is infested by an insect present in economic numbers. The mere fact that an organized program exists has helped to encourage the selective use of insecticides as outlined in Unit 3. Generally the number of insecticide applications per acre has been reduced from levels prior to IPM programs, but increased applications have occurred in a few instances. The primary objective in cotton IPM is not decreased or increased insecticide use, but rather its efficient use. At this time, insecticides are the only quick means of reducing a cotton insect population which has reached or exceeded the economic level.

In the Southwest, insecticide treatment of safflower fields located near cotton is used to prevent lygus migration into cotton as the safflower matures. One insecticide application properly timed to destroy the lygus population in safflower may prevent the necessity of treating the cotton, possibly setting off a chain of events that require several applications.

Other control methods

Several control methods either are being used to a limited extent in cotton IPM or hold promise for increased importance in the future. Resistant varieties for controlling bollworms, boll weevils, lygus bugs, and other insects; pheromone control of pink bollworms and boll weevils by confusion or trapping or both; diapause control of boll weevils by late-season insecticide applications; and the use of sterilized males for controlling boll weevils and pink bollworms are examples of methods being used to a limited extent. With additional development, these and other methods will be added to IPM systems of the future.

Corn

Field corn, totaling 65 million acres, is probably raised in some quantity in every state. It comprises 77 percent of all grains fed to livestock and is obviously an important staple in our agricultural economy (Table 5.1). The following discussion of corn IPM is derived solely from the work being conducted in the corn belt.

There are areas in this region where severe losses are caused by corn insect pests, while in the same year other areas may not be affected by insect attack. This is due mainly to the differences in weather conditions which commonly affect insect populations through the North Central United States. These variations in climate, and in resulting waves in densities of pests which have few or no parasites, predators, or disease pathogens, led to the general practice of preventive insecticide treatments for the control of the soil insect complex.

Many of the problems in corn center on the fact that a number of the insect pests favor other crops in the system over corn, and when corn follows one of these crops it may be severely damaged by the existing pest population which developed in the preceding crop. In Iowa continuous corn and corn following sod result in serious insect problems. This makes crop rotation an all-important part of the corn IPM program.

Several basic lessons emerge from corn IPM: Both the early planting of corn on hay crop ground and no-tillage cultural practices favor damage by seed-corn beetles, seed-corn maggots, and wireworms; white grubs can become a serious problem when corn follows soybeans or hay crops, or when it is planted in no-tillage land; corn root aphids are more severe following grasses or weeds; infestations of grape colaspis follow clover crops; corn rootworm is a problem on continuous corn; black cutworms can be damaging to corn following soybeans if the soybean field is not plowed; billbugs are found in grassy areas; first generation European corn borers survive better in early-planted corn; the southwestern corn borer, the corn earworm, the fall armyworm, and the true armyworm are more of a problem to late-planted corn; true armyworms are more severe in weedy corn and corn on no-tillage fields; chinch bugs and corn leaf aphids are more damaging to late corn on pasture sods or no-tillage fields; and high nitrogen promotes aphid infestation but deters chinch bugs.

Thus, owing to the general lack of parasites, predators, and disease pathogens of the major corn pests, IPM has had to rely heavily on cultural practices to reduce the damage potential of many pest species. The most important of these practices are

1. early planting, which reduces the effects of several species, enhancing only the European corn borer, and occasionally cutworms;
2. using corn varieties which are resistant to the European corn borer and which reduce significantly first-brood problems;
3. reducing irrigation, where used, to reduce survival of corn rootworm larvae after hatching;
4. early harvesting of late-planted corn to reduce stalk breakage and ear droppage resulting from second-brood corn borer damage, and to reduce corn rootworm oviposition;

5. using only corn and soybeans in crop rotation to avoid the wireworm, seed beetle, seed maggot, and armyworm problems resulting from three-way rotations;
6. avoiding insecticide applications to hay crops except when absolutely necessary to retain "nurseries" of parasites and predators.

5.9 Corn IPM systems have had to rely on ____________ control, owing to the lack of parasites, predators, and disease pathogens.

\- - - - - - - - - - - - - - - -

cultural

Economic levels have been developed for first-brood European corn borers, corn leaf aphids, armyworms, and black cutworms, but, surprisingly, none have been developed for the one insect (the corn rootworm) responsible for more soil insecticide use than all the others. Insect population sampling and development of economic levels have not reached the point where large-scale IPM programs are successful in corn. Present techniques place corn IPM on the threshold of area-wide acceptance.

5.10 Economic thresholds have not yet been developed for the ____________, which is primarily responsible for the heavy use of soil insecticides in corn production.

\- - - - - - - - - - - - - - - -

corn rootworm

Alfalfa

Alfalfa, like corn, is probably grown in every state; it is also the fourth largest crop in acreage (Table 5.1) in the United States. This crop is an ideal host for many insect pests, which in turn support a wide variety of predators and parasites. This ongoing nursery of predators and parasites helps control pests not only of alfalfa but of other nearby crops as well. For this reason alone, it is important that insecticide treatment of alfalfa fields be avoided if possible, to preserve this source of beneficial insects.

5.11 Alfalfa fields, if left untreated, serve as ____________ of predators and parasites that migrate to other crops.

\- - - - - - - - - - - - - - - -

- - - - - - - - - - - - - - - - -

nurseries

The key insect pests of alfalfa, depending on geographical area, are the alfalfa weevil, the Egyptian alfalfa weevil, the potato leafhopper, the spittle bug, the spotted alfalfa aphid, the lygus bug, the alfalfa caterpillar, the beet armyworm, the corn earworm, and many others.

The IPM program in alfalfa utilizes all the basic elements of IPM discussed in Unit 2 and many of the components listed in Unit 3. Of the basic elements, natural control is utilized by relying on beneficial insects to control certain pests and by obtaining natural mortality of others through cutting practices, and the like. Sampling to determine economic levels is used to prevent unnecessary insecticide applications, and the biology and ecology of all insects is used to make all management decisions. Several components of IPM are routinely involved with alfalfa production. These are resistant varieties and biological, cultural, and chemical controls.

Indiana has a unique and very advanced alfalfa IPM program which includes the use of local computer terminals that provide customized pest-control recommendations and constantly updated weather reports. The computer terminals are located in area county agricultural agents' offices and are tied in with the central computer at Purdue University. From these terminals, cooperating growers can receive on a moment's notice (1) descriptions of the current insect pest situation and weekly updated recommendations for each area, (2) estimates of plant height, yield, and relative number of various stages of important insects, (3) specific weather conditions and a five-day agricultural forecast as well as an extended thirty-day outlook (also included is a hay forecast giving the expected hay-curing conditons), (4) approved insecticides with their rates and brief application instructions, and (5) brief fertilizer recommendations which are particularly appropriate immediately following a harvest. There are three ways that a grower can use this futuristic system: (1) he can visit the county agent's office and have the report printed from the terminal, (2) he may phone the agent's office and have the report read to him, or (3) he can have the updated reports mailed to him at critical times during the growing season.

The Indiana system represents the best in scientific and agricultural technology and contains those advanced ideas that will become commonplace in IPM technology in the next decade.

Tobacco

The tobacco IPM program probably resembles, more than any other, the system currently used in cotton, because of the use of scouts in sampling pest levels and the past history of insecticide-use patterns. After the introduction of DDT and TDE, insecticide-use patterns consisted of regularly timed insurance applications or applications based purely on intuition. And as a result beneficial species were unduly suppressed.

The important economic insect pests of tobacco are the tobacco hornworm, the tomato hornworm, the tobacco budworm, the corn earworm, and the tobacco flea beetle. The cabbage looper and the green peach aphid are minor pests that can be elevated to major importance following insecticidal destruction of their natural enemies.

The tobacco IPM system that has been developed, and is slowly being accepted by growers, involves the four basic elements of IPM given in Unit 2 on an area-wide basis, and at least two of the components in Unit 3. These are (1) setting of economic levels for the important pests; (2) developing a weekly sampling or scouting system to determine the levels of pests and their parasites and predators; (3) permitting maximum natural suppression of pests by beneficial species; (4) applying selected insecticides only when pest population levels exceed the economic level; (5) carrying out effective sucker control through the properly timed application of growth inhibitors; and (6) conducting thorough stalk destruction after harvest to reduce the late-season food supply used in the buildup of overwintering pest populations. Such an approach provides effective area-wide pressure to the general pest population yet avoids undue suppression of beneficial insects.

5.12 The two key elements that continually emerge in these various IPM systems are good ______________ methods and the setting of ____________________.

- - - - - - - - - - - - - - - -

sampling economic levels

Grain Sorghum

In the past decade grain sorghum has become second to corn as a source of carbohydrates in the diets of all livestock, and large acreages, amounting to 14 million acres, are grown in all of the southern and western states. The primary pests of grain sorghum are the sorghum midge, the greenbug, the corn leaf aphid, the Banks grass mite, the sugarcane borer, the chinch bug, the corn earworm, the fall armyworm, the lesser cornstalk borer, the southwestern corn borer, the stink bug, and the whitefly.

Pest-management programs for sorghum have included (1) early, uniform planting within a community or area to prevent damage by the sorghum midge (egg laying occurs during the brief blooming stage), (2) planting varieties that are resistant to the corn leaf aphid, the sugarcane borer, or the greenbug, (3) rotating crops to avoid wireworm and white grub damage, (4) applying granular insecticide at the minimum rate in the whorl for southwestern corn borer control and to avoid destruction of beneficial insects, (5) applying only one-half the recommended rates of organophosphate insecticides for aphid control to preserve beneficial insects, and (6) establishing economic injury levels for the midge, the greenbug, the corn leaf aphid, and various borers.

5.13 In the grain sorghum IPM system, how are heavy aphid infestations controlled without harming the beneficial insects?

- - - - - - - - - - - - - - -

By applying only one-half the recommended rates of the broad-spectrum organophosphate insecticides.

The introduction of new hybrid varieties, especially the full-season hybrids, and improved cultural practices have doubled the average yields per acre over the past decade. It is not uncommon to obtain 4,000 pounds of grain per acre in the irrigated western states. The complex of insects listed can reduce yields to less than 1,000 pounds per acre if left uncontrolled. For instance, in Arizona the control of the southwestern corn borer alone with a single, precision application of granular insecticide in the whorls, increased yields from 2,000 pounds to 4,000 pounds per acre. It should be kept in mind that the insects listed do not all occur in a single area, but represent the total range of pests throughout the nation. At any given location usually no more than three or four pests would be of economic importance.

With yields of 2 tons per acre, grain sorghum is approaching the importance of wheat in the nation's agriculture and merits the development of intensive IPM programs where trained personnel can provide the necessary supervision.

Peanuts

Insects of economic importance in the production of peanuts fall into the classifications of foliage-feeding insects and soil insects. The problems and methods of dealing with these two groups are quite different.

The soil insects feed primarily on the peanut beneath the soil surface. The most important in this group are the lesser cornstalk borer and the southern corn rootworm. Economic thresholds have not been developed for either insect, and since control is confined to insecticides applied to the soil, treatment normally begins at the first signs of infestation. Cultural control of these pests is not satisfactory. Infestations of the lesser cornstalk borer can be reduced by avoiding soil moisture stress with adequate and properly timed irrigations. The Southern corn rootworm, in contrast, tends to build up during extended periods of excess soil moisture.

Foliage feeders include the corn earworm, the granulate cutworm, the fall armyworm, the beet armyworm, the velvet bean caterpillar, and the cabbage looper. As with most crops that sustain varying levels of foliage feeding, the peanut plant can endure the loss of considerable foliage with no apparent economic loss. Georgia research has shown that four caterpillars per linear foot of row in any combination represent an economic level. Weekly inspections of fields are used to determine infestation levels, and insecticidal control is applied as needed. Peanut IPM programs in Texas and Oklahoma have gone to complete programs which include weeds, diseases, and nematodes. As you would guess by now, however, management of any pest includes two essentials: economic levels and an appropriate sampling system. Given these two tools, decision making in insecticide used by the growers is relatively free of guess and gamble.

Citrus

The citrus-producing states, in alphabetical order, are Arizona, California, Florida, Louisiana, and Texas. In In each state, except Arizona, the primary pests are several species of the spider mite and the scale insect. The citrus thrips is the main pest in Arizona and is also important in California. Other pests that sometimes require control in certain areas include mealy bugs, white flies, aphids, katydids, plant bugs and stink bugs.

Unfortunately, as is the case in the vegetable industry, much of the control applications for pest species is for the purpose of improving the cosmetic quality of the product, presumably aimed at pleasing the consumer. When cosmetic quality is the basis for control measures, the economic threshold is essentially zero. Consequently most of the pests of importance are controlled when they appear rather than when they reach an economic level.

California has progressed into citrus IPM more than the other citrus-growing states combined. There, growers in three southern counties formed citrus-pest-control co-ops for using biological and chemical controls. The parasites of mealy bugs and scales were mass-reared by the co-op supervisors who were responsible for the release of these beneficial insects as well as for selecting and applying pesticides that would cause them the least harm. Because of their beneficial insect-salvaging quality, petroleum oil applications are still extensively used in the coastal areas instead of the more effective broad-spectrum pesticides. IPM in varying degrees has been in effect since the early 1930's in southern California orchards because pest-control supervisors have usually been individuals with both the education and the experience to recognize the beneficial insects as well as the pest species.

Vegetables

The presence of insects, insect parts, and insect frass is not tolerated in fresh or processed vegetables by either the Food and Drug Administration or the consuming public. This results in near-zero economic levels and makes vegetable IPM the most challenging of management systems. There is a total lack of economic levels of pests affecting most vegetable crops. Unfortunately, in most situations the presence of any insects on the crops means economic consequence and possible failure to the grower. Obviously it appears that such situations are locked in position with little hope of relief.

A good example of an emerging IPM system is a two-year study on lettuce in Arizona. Historically, fall lettuce is the victim of heavy worm infestations comprised mainly of cabbage loopers, beet armyworms, and corn earworms. Growers have resorted to insecticide applications two to three days apart at costs approaching $200 per acre in attempting to keep the crop free of these insects. The first step was to develop a sampling method and to answer the question of how many worms could be tolerated without affecting the number or quality of marketable heads. The regularly timed applications, at four- to six-day intervals, averaged 14.5 for the two seasons. Other blocks of lettuce were permitted to maintain worm populations from one to twenty infested heads per twenty sampled. Grading by the standard marketing system showed that up to three heads could be infested with no change in quantity or quality of marketable heads. This level of infestation was maintained by using only 5 insecticide applications for the season compared with the average 14.5 used to provide continuous insecticide protection.

This is a sound beginning to solve a difficult problem. The two essential basic ingredients, sampling method and economic injury levels, were developed to serve as a basis for control decisions. When these two conditions are met, an IPM system can be developed on any crop under any circumstances.

Tree Fruit

The development of IPM programs on tree fruit (apples, pears, peaches, cherries, and so on) has been hampered by inadequate sampling methods and lack of good economic levels. Much work has been devoted to the IPM of tree-fruit insect pests, particularly in the states of Washington, Michigan, Pennsylvania, and New York. In the late 1960's it was well confirmed that the complete removal of pesticides is usually disastrous, since natural control of tree-fruit pests, especially the codling moth, the red-banded leaf roller, the plum curculio, the San Jose scale, the apple maggot, the pear psylla, and the Oriental fruit moth, is almost negligible.

Lead arsenate was used heavily from 1900 to 1946, and the first half of the twentieth century can be termed as the lead arsenate period. DDT rapidly replaced lead arsenate after World War II and was used extensively from 1946 to 1960. By then the codling moth, the red-banded leaf roller, and the leafhopper had developed resistance, and mites were resistant to most available acaricides.

The resistant-insect problems were solved by switching from the organochlorine insecticides to the organophosphates and the carbamates. Presently all major and most minor pests are handled by available pesticides. However, good substitutes are not available, and serious problems could develop when resistance to organophosphates and carbamates develops. Though the key insect pests are readily controlled, all the pest species of mites have developed resistance to available acaricides. Consequently mite control is a continually changing problem.

With respect to IPM, tree-fruit pests consist of fruit feeders and nonfruit feeders. The economic levels for fruit feeders are essentially zero for commercial fruit production because of the cosmetic standards of wholesale buyers and consumers, and because of loss of production by growers. The nonfruit feeders can be tolerated in fairly high numbers without measurable damage or economic loss. This group is composed mainly of mites and occasionally leafhoppers, and it is in controlling this group that IPM has its greatest potential.

The greatest advancements to date have been the reduction of acaricide use and the return of predaceous mites into the system. Assessing pest populations with pheromones for the codling moth, the red-banded leaf roller, and the Oriental fruit moth, is still in the experimental but highly promising stage.

Light traps have been useful in determining moth activity, but are likely to be replaced by easier to handle and more selective pheromone traps as they become available. Trapping techniques offer good possibilities for reducing the number of pesticide applications through improved timing of sprays and better monitoring of pest populations. Mass trapping or prediction of infestations of fruit-feeding pests are, at best, several years away.

Additional inputs for tree-fruit IPM are the use of dormant sprays applied when few parasites and predators are active; the avoidance of spraying the ground-cover crops which serve as good sources of beneficial insects; spot-treatment in orchards, applying pesticides only where the pest exists; the use of selective acaricides which preserve predators; and soil application of insecticides for control of plum curculio. The integration of chemical and biological controls of deciduous tree-fruit pests has been attained mostly by using broad-spectrum insecticides in ways that minimize their effects on predators and parasites.

A pest-management program on pears in Oregon has demonstrated that several species of mites are controlled by a predaceous (phytoseiid) mite. However, by controlling the pest species with acaricides, the predators were also suppressed. Predator mite populations finally increased to effective numbers when the insecticidal rate for the summer control of the codling moth was reduced to a very low level.

Patience is required in programs, as is demonstrated here. The changeover from a standard program, which included several applications of acaricides, to the management program resulted in economic injury from mites the first year; however, in the second and third years the predaceous mites held the pest mites below injury levels.

Generally speaking, tree-fruit IPM currently depends on the heavy use of pesticides for key pest and most mite control. Natural control plays only a minor role except in the case of mites, where IPM has made the most progress. Economic levels are zero for the important pests, and monitoring these pests is difficult, since simplified, proven, and accepted pheromone devices have not become available. Until improved sampling methods for the major pests are developed, dependence on the heavy use of chemicals through timed spray schedules will continue.

Soybeans

Soybean IPM programs are just getting under way, and these systems, when fully developed, should have a real impact on both yields and production costs. The primary problems are lack of sound economic levels for major pests and inadequate sampling methods.

The major pests of soybeans are the corn earworm, the Mexican bean beetle, the green cloverworm, the soybean looper, the velvet bean caterpillar, the three-cornered alfalfa hopper, and stink bugs. There are three stages of development in this crop which require changes in economic levels: prebloom, bloom, and pod-filling. Damage to the plant is of four basic types: foliage feeding, blossom injury, feeding on the developing pods, and puncture of the beans through pods.

Soybean IPM strategies are being developed in Arkansas, Florida, Illinois, Kentucky, Indiana, Iowa, Louisiana, Mississippi, Missouri, North Carolina, and South Carolina. Of greatest importance in these programs are developing practical survey methods and economic levels for the key pests; using minimum rates of insecticides to reduce pest populations without destroying the beneficial populations; planting of resistant varieties; interplanting of other crops, such as alfalfa, in a strip-cropping fashion to provide a source of predators and parasites; and timing of the appropriate insecticide to achieve maximum pest reduction.

With the wide range of geographical areas in which soybeans are grown, and the wide variety of pests that inhabit these areas, IPM systems developed in one area may be totally unacceptable in anothers. Soybean IPM strategies, like those for vegetables and tree fruit, are just emerging into useful systems and will certainly be accepted by commercial growers when their usefulness and economy are demonstrated.

References

Anonymous. 1973. Alfalfa pest management in Indiana. A report on progress for the first year. Coop. Ext. Serv., Purdue University, West Lafayette, Ind. 26 pp.

Barnes, G., L. Moore, C. F. Sarter, and L. M. Sparks. 1972. Present status of cotton insect pest management. Implementing practical pest management strategies: Proceedings of a National Extension Insect-Pest Management Workshop. Purdue University, West Lafayette, Ind. Mar. 14-16.

Deal, A. S. and J. E. Brogdon. 1972. Present status of pest management in citrus. Ibid.

Ellis, H. C. 1974. Tobacco insect pest management. Ibid.

Ganyard, M. C., Jr., and H. C. Ellis. 1972. A tobacco pest management pilot project in North Carolina. Ibid.

Glass, E. H. and S. C. Hoyt. 1972. Insect and mite pest management on apples. Ibid.

Godfrey, G. L. (Editor) 1974. Selected literature of soybean entomology. International Agriculture Publications, INTSOY Series No. 1. University of Illinois, Urbana. 224 pp.

Henning, R. J. 1974. Peanut pest management in Georgia, a comprehensive approach. Implementing practical pest management strategies: Proceedings of a National Extension Insect-Pest Management Workshop. Purdue University, West Lafayette, Ind. Mar. 14-16.

Hines, B. M. and C. E. Hoelscher. 1974. Progress report of 1973 Texas peanut pest management. Peanut and Tobacco Pest Management Workshop Proceedings. Coop. Ext. Serv., Okla. State University, Stillwater, Okla., Feb. 14-15.

Hoyt, S. C. and E. C. Burts. 1974. Integrated control of fruit pests. Ann. Rev. Entomol. 19:231-252.

Jeppson, L. R. 1974. Pest management in citrus orchards. Bull. Entomol. Soc. Amer. 20(3):221-222.

Petty, K. B. and R. T. Huber. 1972. Corn insect pest management. Implementing practical pest management strategies: Proceedings of a National Extension Insect-Pest Management Workshop. Purdue University, West Lafayette, Ind. Mar. 14-16

Research on grain sorghum insects and spider mites in Texas. 1971. Consolidated PR-2863-2876 (January). Texas A&M Univ. Coll. Sta., Tex.

Sturgeon, R. V., Jr. 1974. Annual report of 1973 peanut pest management program (Oklahoma). Peanut and Tobacco Pest Management Workshop Proceedings. Coop. Ext. Serv., Okla. State University, Stillwater, Okla., Feb. 14-15.

Westigard, Peter H. 1971. Integrated control of spider mites on pear. J. Econ. Entomol. 64(2):496-501.

APPENDIX

Common and Scientific Names of Specific Insects and Mites Mentioned in the Text

alfalfa caterpillar	*Colias eurytheme* Boisduval
alfalfa looper	*Autographa californica* (Speyer)
alfalfa weevil	*Hypera postica* (Gyllenhal)
apple maggot	*Rhagoletis pomonella* (Walsh)
armyworm	*Pseudaletia unipuncta* (Haworth)
Banks grass mite	*Oligonychus pratensis* (Banks)
beet armyworm	*Spodoptera exigua* (Hübner)
beet leafhopper	*Circulifer tenellus* (Baker)
big-eyed bug	*Geocoris punctipes* Say
black cutworm	*Agrotis ipsilon* (Hufnagel)
boll weevil	*Anthonomus grandis* Boheman
bollworm	*Heliothis zea* (Boddie)
cabbage looper	*Trichoplusia ni* (Hübner)
catalpa sphinx	*Ceratomia catalpae* (Boisduval)
chinch bug	*Blissus leucopterus* (Say)
citrus thrips	*Scirtothrips citri* (Moulton)
codling moth	*Laspeyresia pomonella* (L.)
corn earworm	*Heliothis zea* (Boddie)
corn leaf aphid	*Rhopalosiphum maidis* (Fitch)
cotton aphid	*Aphis gossypii* Glover
cotton leafperforator	*Bucculatrix thurberiella* Busck
cotton leafworm	*Alabama argillacae* (Hübner)
cottony-cushion scale	*Icerya purchasi* Maskell
cyclamen mite	*Steneotarsonemus pallidus* (Banks)
Egyptian alfalfa weevil	*Hypera brunneipennis* (Boheman)
European chafer	*Amphimallon majalis* (Razoumowdky)
European corn borer	*Ostrinia nubilalis* (Hübner)
fall armyworm	*Spodoptera frugiperda* (J. E. Smith)
floodwater mosquito	*Aedes sticticus* (Meigen)
granulate cutworm	*Feltia subterranea* (Fabricius)
grape colaspis	*Colaspis flavida* (Say)
grape leafhopper	*Erythroneura elegantula* (Osburn)
grape phylloxera	*Phylloxera vitifoliae* (Fitch)
greenbug	*Schizaphis graminum* (Rondani)
green cloverworm	*Plathypena scabra* (Fabricius)
green peach aphid	*Myzus persicae* (Sulzer)
gypsy moth	*Porthetria dispar* (L.)
Hessian fly	*Mayetiola destructor* (Say)
hickory shuckworm	*Laspeyresia caryana* (Fitch)
honey bee	*Apis mellifera* Linnaeus
house fly	*Musca domestica* (L.)
Japanese beetle	*Popillia japonica* Newman
lesser cornstalk borer	*Elasmopalpus lignosellus* (Zeller)
lygus bug	*Lygus* spp.

McDaniel mite	*Tetranychus mcdanieli* McGregor
meadow spittlebug	*Philalmus spumorius* (L.)
Mediterranean fruit fly	*Ceratitis capitata* (Wiedemann)
melon fly	*Dacus cucurbitae* Coquillett
Mexican bean beetle	*Epilachna varivestis* Mulsant
minute pirate bug	*Orius insidiousis* (Say)
Nantucket pine tip moth	*Rhyacionia frustrana* (Comstock)
northern corn rootworm	*Diabrotica longicornis* (Say)
oriental fruit fly	*Dacus dorsalis* Hendel
oriental fruit moth	*Grapholitha molesta* (Busck)
pea aphid	*Acyrthosiphon pisum* (Harris)
pear psylla	*Psylla pyricola* Foerster
pink bollworm	*Pectinophora gossypiella* (Saunders)
plum curculio	*Conotrachelus nenuphar* (Herbst)
potato leafhopper	*Empoasca fabae* (Harris)
potato tuberworm	*Phthorimaea operculella* (Zeller)
red-banded leaf roller	*Argyrotaenia velutinana* (Walker)
rhinoceros beetle	*Xyloryctes jamaicensis* (Drury)
rice water weevil	*Lissorhoptrus oryzophilus* Kuschel
salt-marsh caterpillar	*Estigmene acrea* (Drury)
San Jose scale	*Aspidiotus perniciosus* Comstock
screwworm	*Cochliomyia hominivorax* (Coquerel)
seed-corn beetle	*Agonoderus lecontei* Chaudoir
seed-corn maggot	*Hylemya platura* (Meigen)
sorghum midge	*Contarinia sorghicola* (Coquillett)
southern cornstalk borer	*Diatraea crambidoides* (Grote)
southern corn rootworm	*Diabrotica undecimpunctata howardi* Barber
spotted alfalfa aphid	*Theriophis trifolii* (Monell)
sugarcane borer	*Diatraea saccharalis* (Fabricius)
sweet potato weevil	*Cylas formicarius elegantulus* (Summers)
three-cornered alfalfa hopper	*Spissistilus festinus* (Say)
tobacco budworm	*Heliothis virescens* (Fabricius)
tobacco flea beetle	*Epitrix hirtipennis* (Melsheimer)
tobacco hornworm	*Manduca sexta* (Johannson)
tomato hornworm	*Manduca quinquemaculata* (Haworth)
two-spotted spider mite	*Tetranychus urticae* (Koch)
vedalia beetle	*Rodolia cardinalis* (Mulsant)
velvetbean caterpillar	*Anticarsia gemmatalis* Hübner
wheat curl mite	*Aceria tulipae* (Keifer)
wheat stem sawfly	*Cephus cinctus* Norton

GLOSSARY

Acetylcholine (ACh) — Chemical transmitter of nerve and nerve-muscle impulses in animals.

Adjuvant — An ingredient that improves the properties of a pesticide formulation. Includes wetting agents, spreaders, emulsifiers, dispersing agents, foam suppressants, penetrants, and correctives.

Agroecosystem — An agricultural area sufficiently large to permit long-term interactions of all the living organisms and their nonliving environment.

Alkylating agent — Highly active compounds that replace hydrogen atoms with alkyl groups, usually in cells undergoing division.

Antibiosis — The tendency of a plant to resist insect injury by preventing, injuring or destroying insect life.

Antimetabolite — Chemicals that are structurally similar to biologically active metabolites, and which may take their place detrimentally in a biological reaction.

Applied ecology — The practice of utilizing natural control of insect pests to the fullest extent. Naturally-occurring biological control is applied ecology.

Attractant, insect — A substance that lures insects to trap or poison-bait stations. Usually classed as food, oviposition, and sex attractants.

Augmentation — The practice of assisting biological control agents such as predators and parasites to make them more effective.

Aziridine — A chemical classification of chemosterilants containing 3-membered rings composed of one nitrogen and two carbon atoms.

Balance — A status of insect populations whereby large deviations from population oscillations do not occur.

Beneficial insect — An insect that serves the best interests of man, such as insect predators and parasites or pollinating insects.

Biological control agent — Any biological agent that adversely affects pest species.

Biomagnification — The increase in concentration of a pollutant in animals as related to their position in a food chain, usually referring to the persistent, organochlorine insecticides and their metabolites.

Biotic insecticide — Usually microorganisms known as insect pathogens that are applied in the same manner as conventional insecticides to control pest species.

Biotype — A population or group of individuals composed of a single genotype.

Broad-spectrum insecticide — Nonselective, having about the same toxicity to most insects.

Carcinogen — A substance that causes cancer in living animal tissue.

Carbamate insecticide — One of a class of insecticides derived from carbamic acid.

Carrier — An inert material that serves as a diluent or vehicle for the active ingredient or toxicant.

Certified applicator — Commercial or private applicator qualified to apply restricted-use pesticides as defined by the EPA.

Chemosterilant — Chemical compounds that cause sterilization or prevent effective reproduction.

CHEMTREC — A toll-free, long-distance, telephone service that provides 24-hr emergency pesticide information. (800-424-9300).

Cholinesterase (ChE) — An enzyme of the body necessary for proper nerve function that is inhibited or damaged by organophosphate or carbamate insecticides taken into the body by any route.

Colonization — The introduction and establishment of a species in a new area, usually referring to beneficial insects.

Compatible — (Compatibility) When two materials can be mixed together with neither affecting the action of the other.

Concentration — Content of a pesticide in a liquid or dust, for example lbs/gallon or percent by weight.

Cosmetic quality — Eye-appealing characteristics of a fruit or vegetable which have no relationship with nutritional quality, e.g. color, no blemishes.

Cotton square — Fruiting bud of the cotton plant.

Crop diversification — Cropping system where a number of different crops are planted in the same general area and may be rotated from field to field, year after year.

Dermal toxicity — Toxicity of a material as tested on the skin, usually on the shaved belly of a rabbit.

Diapause — A physiological state of arrested development, generally resulting from physical stimuli, such as temperature and light, that provides the insect a means of surviving unfavorable periods.

Diluent — Component of a dust or spray that dilutes the active ingredient.

DNA — Deoxyribonucleic acid.

Dose, dosage — Same as rate. The amount of toxicant given or applied per unit of plant, animal, or surface.

Drift — Movement of airborne pesticide from the intended area of application.

Dynamic economic level — A term used to denote a changing relationship of a pest insect and its host plant, indicating that different levels of the pest are required to cause damage depending on the stage of growth or season.

Ecdysone — Hormone secreted by insects essential to the process of molting from one stage to the next.

Ecology — The science dealing with the relationship of living organisms and their nonliving environment.

Economic injury level — The insect pest level at which economic injury occurs.

Economic level — The insect pest level at which additional management practices must be employed to prevent economic losses.

Ecosystem — The interacting system of all the living organisms of an area and their nonliving environment.

Electromagnetic spectrum — The entire wave-form energy spectrum including radio waves, light, and X- and gamma-rays.

Emulsifier — Surface active substances used to stabilize suspensions of one liquid in another, for example, oil in water.

Emulsion — Suspension of miniscule droplets of one liquid in another.

EPA — The Environmental Protection Agency

Exterminate — Often used to imply the complete extinction of a species over a large continuous area such as an island or a continent.

Fecundity — Reproductive capacity.

FEPCA — Federal Environmental Pesticide Control Act of 1972.

Field scout — A person who samples fields for insect infestations.

FIFRA — Federal Insecticide, Fungicide and Rodenticide Act of 1947.

Formamidine insecticide — A new group with a new mode of action highly effective against insect eggs and mites.

Formulation — The form in which a pesticide is sold for use.

Fumigant — A volatile material that forms vapors which destroy insects, pathogens, and other pests.

Gallonage — The number of gallons of finished spray mix applied per acre, tree, hectare, square mile, or other unit.

Harvest intervals — The period between last application of a pesticide to a crop and the harvest as permitted by law.

Hormone — A product of living cells that circulates in the animal or plant fluids and that produces a specific effect on cell activity remote from its point of origin.

Host-plant resistance — Inherited qualities of plants that influence the extent of insect damage.

Hydrolysis — Chemical process of (in this case) pesticide breakdown or decomposition involving a splitting of the molecule and addition of a water molecule.

Insect-growth regulator (IGR) — Chemical substance which disrupts the action of insect hormones controlling molting, maturity from pupal stage to adult, and others.

Insect pest management — The practical manipulation of insect (or mite) pest populations using any or all control methods in a sound ecological manner.
Integrated control — The integration of the chemical and biological control methods.
Interplanting — The interplanting of one crop within another for the purpose of trapping pest insects.
Key pest — An insect that is routinely present sometime during the growing season and causes economic damage.
LC_{50} — A lethal concentration of a substance in air or liquid for 50% of the test organisms.
LD_{50} — A lethal dose for 50% of the test organisms. The dose of a toxicant producing 50% mortality in a population. A value used in presenting mammalian toxicity, usually oral toxicity, expressed as milligrams of toxicant per kilogram of body weight (mg/kg).
Microbial insecticide — A microorganism applied in the same way as conventional insecticides to control an existing pest population.
Mutagen — Substance causing genes in an organism to mutate or change.
Oral toxicity — Toxicity of a compound when given by mouth. Usually expressed as number of milligrams of chemical per kilogram of body weight of animal (white rat) when given orally in a single dose that kills 50% of the animals. The smaller the number, the greater the toxicity.
Organochlorine insecticide — One of the many chlorinated insecticides, e.g. DDT, dieldrin, chlordane, BHC, Lindane, etc.
Organophosphate — Class of insecticides derived from phosphoric acid esters.
Ovicide — A chemical that destroys an organism's eggs.
Oviposition — The act of laying or depositing eggs.
Parasite — In this book it refers to insects that develop in or on another insect, eventually killing the host.
Pathogen — Any disease-producing organism or virus.
Periodic colonization — The periodic release of parasites or predators to reestablish a favorable ratio of pests and beneficial insects.
Persistence — The quality of an insecticide to persist as an effective residue due to its low volatility and chemical stability, e.g. certain organochlorine insecticides.
Pesticide — An "economic poison" defined in most state and federal laws as any substance used for controlling, preventing, destroying, repelling, or mitigating any pest. Includes fungicides, herbicides, insecticides, nematicides, rodenticides, desiccants, defoliants, plant growth regulators, etc.
Pest-management specialist — A person with sufficient training and experience to successfully manage insect populations.

Pheromones — Highly potent insect sex attractants produced by insects. For some species laboratory-synthesized pheromones have been developed for trapping purposes.

Physical selectivity — Refers to the use of broad-spectrum insecticides in such ways as to obtain selective action. This may be accomplished by timing, dosage, formulation, etc.

Physiological selectivity — Refers to insecticides which are inherently more toxic to some insects than to others.

Phytotoxic — Injurious to plants

Point sample — A method of sampling for insects designed to relate the number of insects or their damage to the number of plants and/or plant parts per acre.

Poison Control Center — Information source for human poisoning cases, including pesticide poisoning, usually located at major hospitals.

Predator — Relative to insects, a predator is an insect that feeds on others, usually requiring several to complete development.

Preference — (or nonpreference) Plant characters and insect responses that lead to or away from the use of a particular plant or variety for oviposition, food, or shelter.

Protective clothing — Clothing to be worn in pesticide-treated fields under certain conditions as required by federal law, e.g. reentry intervals.

Random sampling — The most commonly used method of sampling for insects whereby samples are taken at random, with good field coverage, to determine insect numbers or damage.

Raw agricultural commodity — Any food in its raw and natural state, including fruits, vegetables, nuts, eggs, raw milk, and meats.

Reentry (intervals) — That waiting interval required by federal law between application of certain hazardous pesticides to crops and the entrance of workers into those crops without protective clothing.

Repellent (insects) — Substance used to repel ticks, chiggers, gnats, flies, mosquitoes, and fleas.

Residue — Trace of a pesticide and its metabolites remaining on and in a crop, soil, or water.

Resistance (insecticide) — Natural or genetic ability of an organism to tolerate the poisonous effects of a toxicant.

Restricted-use pesticide — one of several pesticides, designated by the EPA, that can be applied only by certified applicators, because of their inherent toxicity or potential hazard to the environment.

RNA — Ribonucleic acid.

Safener — Chemical that reduces the phytotoxicity of another chemical.

Secondary pest — A pest which usually does little if any damage but can become a serious pest under certain conditions, e.g. when insecticide applications destroy its predators and parasites.
Selective insecticide — One which kills selected insects, but spares many or most of the other organisms, including beneficial species, either through differential toxic action or the manner in which insecticide is used.
Sequential sampling — A method of sampling for insects that requires continued sampling until a preestablished upper or lower infestation level is found.
Sex lure — A synthetic chemical which acts as the natural lure (pheromone) for one sex of an insect species.
Spreader — Ingredient added to spray mixture to improve contact between pesticide and plant surface.
Strip-cutting — A term used to denote the practice of harvesting alternate borders of alfalfa so that partially-grown alfalfa is maintained in the field at all times.
Suicidal emergence — Refers to emergence of insects at a time when host plants are unavailable for reproduction and the emerging insects die without reproducing.
Surfactant — Ingredient that aids or enhances the surface-modifying properties of a pesticide formulation (wetting agent, emulsifier, spreader).
Suspension — Finely divided solid particles dispersed in a liquid.
Synergism — Increased activity resulting from the effect of one chemical on another.
Systemic — Compound that is absorbed and translocated throughout the plant or animal.
Teratogenic — Substance which causes physical birth defects in the offspring following exposure of the pregnant female.
Tolerance (residue) — Amount of pesticide residue permitted by federal regulation to remain on or in a crop. Expressed as parts per million (ppm).
Tolerance (host-plant resistance) — A basis of resistance in which the plant shows an ability to grow and reproduce or repair injury in spite of supporting a population of insects that would damage a susceptible variety.
Toxicant — A material that exhibits toxicity.
Trade name — (Trademark name, proprietary name, brand name) Name given a product by its manufacturer or formulator, distinguishing it as being produced or sold exclusively by that company.
Trap cropping — A method of controlling pest insects by planting a more favored host crop and thereby trapping the insects to prevent their movement into the main crop.
Ultra low volume (ULV) — Sprays that are applied at 0.5 gallon or less per acre or sprays applied as the undiluted formulation.

Ultraviolet (UV) — That portion of the light spectrum beyond the visible violet (to humans) but both visible and highly attractive to nocturnal insects.
Vector — An organism, as an insect, that transmits pathogens to plants or animals.
Wettable powder — Pesticide formulation of toxicant mixed with inert dust and a wetting agent which mixes readily with water and forms a short-term suspension (requires tank agitation).

INDEX